Innovations of Green Process Engineering for Sustainable Energy and Environment 2015

Topical Conference at the 2015 AIChE Annual Meeting

Salt Lake City, Utah, USA
8-13 November 2015

ISBN: 978-1-5108-1852-1

Printed from e-media with permission by:

Curran Associates, Inc.
57 Morehouse Lane
Red Hook, NY 12571

CURRAN ASSOCIATES INC. proceedings.com

Some format issues inherent in the e-media version may also appear in this print version.

Printed by Curran Associates, Inc. (2016)

AIChE
120 Wall Street, FL 23
New York, NY 10005-4020

Phone: (800) 242-4363
Fax: (203) 775-5177

www.aiche.org

Additional copies of this publication are available from:

Curran Associates, Inc.
57 Morehouse Lane
Red Hook, NY 12571 USA
Phone: 845-758-0400
Fax: 845-758-2634
Email: curran@proceedings.com
Web: www.proceedings.com

TABLE OF CONTENTS

437892 Reversible Carbonate & CO2 Reaction System for High-Temperature Thermo-Chemical Energy Storage

Abstract:

This presentation will discuss thermochemical energy storage at 300 – 750°C based on reversible chemical reactions between CO_2 and alkaline metal oxides. These materials are inexpensive and abundant. During energy storage, the carbonate is heated up and decomposed into metal oxides and CO_2. The resulting metal oxide is stored at high temperatures, while the CO_2 gas is stored in an adsorbent vessel at ambient temperature. During energy release, the stored CO_2 reacts with the metal oxide to form the carbonate. The heat of chemical reaction involved in the process is several times higher than the typical latent heat provided by conventional phase change materials. High CO2 sorption capacity and rapid sorption/decomposition kinetics are critical to development of a compact storage system. A phase-transfer catalyzed sorbent system and monolith bed will be discussed in this presentation to dramatically enhance intrinsic reaction kinetics and mass transfer while maintain stable working capacity.

422135 The Solar Flow Battery: A Hybrid Device for Base-Load Solar Electricity

Abstract:

Solar energy is by far the most abundant renewable energy resource; indeed the energy in sunlight that strikes the troposphere exceeds global primary consumption by a factor of ~10,000. After many decades of development, photovoltaic technologies are approaching cost parity with fossil-derived electricity. However, the temporal intermittency of sunlight prevents photovoltaics from scaling indefinitely while maintaining a stable electrical grid. Thus there is a critical need for efficient and low-cost energy storage to enable dispatchable solar electricity. To this end, significant advances have been made in the academic and commercial sectors in solar-driven fuel synthesis as well as large-scale battery technologies. Among the most attractive approaches for grid-scale electricity storage is the redox flow battery (RFB), a flowable galvanic cell that, unlike solid-state batteries, decouples energy and power density and can be optimized for efficiency, cost, and robust operation at a large scale. Although several RFB technologies have achieved limited commercial success, significant advances are needed to prove that RFB technologies can be cost-effective when used with intermittent renewables.

Herein I will present a promising hybrid technology that combines aspects of solar photovoltaics with redox flow batteries, which we call the solar flow battery (SFB). The SFB concept is similar to solar-driven water splitting devices, in that it uses semiconductor light absorbers directly interfaced with chemical species that can undergo uphill redox reactions for net solar energy storage. Unlike solar fuels devices, however, the SFB makes use of electrolytes that remain in the liquid phase and exhibit facile electron-transfer kinetics. I will discuss first the operating principles of the SFB, emphasizing several straightforward design rules and the proposed benefits over alternative energy storage approaches. Then I will describe laboratory-scale demonstration devices that we have built based on a unique earth-abundant semiconductor material, WSe_2, that is attractive for use in SFBs due to its favorable electronic properties and high stability under electrochemical conditions.

422845 Low Temperature Direct Biomass-to-Electricity Fuel Cell

Abstract:

Hereby we present a novel fuel cell that directly consumes natural polymeric biomasses, such as starch, cellulose, lignin, and even switchgrass and wood powders. This fuel cell combines some features of solar cells, fuel cells, and redox flow batteries. Specifically, the polyoxomatelate forms a charge transfer complex with the biomass by absorbing solar energy via either photochemical or solar-thermal effects. The charge transfer complex can then release electrons on the carbon anode. The electrons then pass through an external circuit and are captured by oxygen at the cathode. The power density of the solar-induced hybrid fuel cell powered by cellulose reaches 0.72 mW/cm^2, which is almost 1,000 times higher than cellulose based microbial fuel cells. Unlike most cell technologies that are sensitive to impurities, the cell reported in this study is inert to most organic and inorganic contaminants present in the fuels. The fuel cell is completely noble metal free.

Pressure Tubular Reactor Used in Ethylene Hydroformylation for the Production of Propionaldehyde

Pu Yang, Ling Wang, Yanyan Diao, Lei Wang, Suojiang Zhang*

Beijing Key Laboratory of Ionic Liquids Clean Process, Key Laboratory of Green Process and Engineering, State Key Laboratory of Multiphase Complex Systems, Institute of Process Engineering, Chinese Academy of Sciences, Beijing100190, PR China

*Corresponding author. E-mail address: sjzhang@home.ipe.ac.cn.

Hydroformylation catalyzed by phosphine-modified-rhodium is typically used to convert light olefins to aldehydes, which can be further used in the synthesis of alcohols, acids, esters and so on [1, 2]. Generally, it is an important homogeneous catalytic reaction occurred in a stirred tank reactor at high pressure.

Compared with a stirred tank reactor, the pressurized tubular reactor has some advantages in the respects of cost and operation. Herein, a tubular reactor applied in the synthesis of propionaldehyde from ethylene, CO, and H_2 wss investigated in this work. The reactor system with propionaldehyde production of 1-7 kg per day was independently designed and built based on the previous small scale experimental resutls. The effects of gas flow, temperature, pressure, reaction time on the conversion and product selectivity were studied. Under the reaction conditions of CO : CH_2=CH_2 : H_2=1 : 1.1 : 4 (mole ratio), average temperature of 95 ℃ and 1.95 MPa, as shown in Fig. 1, the conversion of ethylene, the selectivity to propionaldehyde and the TOF was 55%, 99% and 8000 h^{-1} respectively under the stable state running.

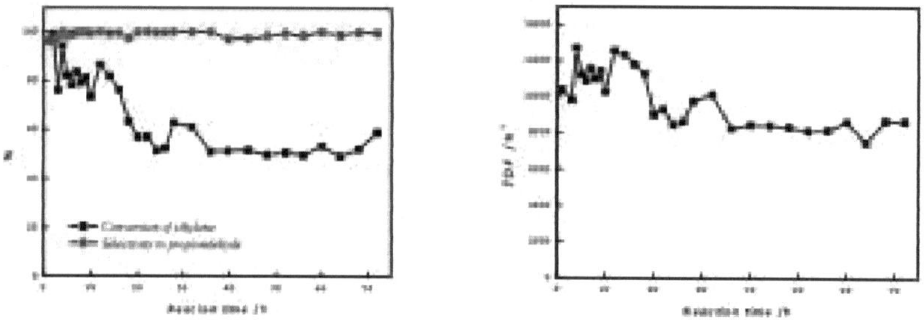

Fig. 1 Conversion of ethylene, selectivity to propionaldehyde and TOF of long term running

Key words: Ethylene, Hydroformylation, Tubular reactor, Production, Propionaldehyde

Acknowledgments

This work was supported by the "Strategic Priority Research Program" of the Chinese Academy of Sciences (XDA07070600).

References

[1] G. Kiss, E.J. Mozeleski, K.C. Nadler, E. VanDriessche, C. DeRoover. *Journal of Molecular Catalysis A: Chemical* 138 (1999) 155–176.

[2] Maizatul S.Shaharun, BinayK.Dutta, Hilmi Mukhtar, Saikat Maitra. *Chemical Engineering Science* 65 (2010) 273–281.

435159 Screening of Metal Oxide Materials for Solar Thermochemical Water Splitting

Abstract:

While solar energy is the most abundant renewable energy resource, the capture, storage, and distribution of it remains a challenge. Solar thermochemical water splitting (STWS) provides a promising route for efficient utilization of this disperse resource since it allows for use of the entire solar spectrum to convert water to an energy dense fuel, H_2. However, despite a significant number of materials having been examined, an optimal redox material to drive this process has yet to be developed. In order to evaluate potential materials, we have developed a set of screening criteria based on thermodynamic requirements. This criteria stipulates that the reduction enthalpy for the material be greater than 286 kJ/mol, which is the overall enthalpy of the water splitting reaction; however, higher reduction enthalpies will also result in lower extents of reduction at a given operating temperature, and therefore less H_2 production. Thus, we evaluate a large set of metal oxide redox materials based on their calculated reduction enthalpies to determine which potential materials are good STWS candidates. In this work we will discuss the findings from our thermodynamic screening of metal oxide materials for solar thermochemical water splitting based on the aforementioned criteria using DFT, as well as our experimental validation of these calculations. We will also briefly describe a method we have developed to evaluate metal oxides for STWS based on their water splitting kinetics.

434922 Cation Exchange in Forward Osmosis: Elucidating Impacts of Membrane and Solution Chemistry

Abstract:

Thin film composite (TFC) membranes designed for use in forward osmosis (FO) have recently reached the ubiquity previously occupied by the cellulose triacetate membranes. FO processes make use of an osmotic pressure gradient to drive water flux across a semi-permeable membrane. The solution having a higher osmotic pressure, or draw solution, can make use of a myriad of different solutes from nanoparticles, to organic solvents, to electrolytes. Electrolyte draw solutions are commonly used in many large scale applications of forward osmosis processes.

A unique interaction between draw and feed solution containing differing electrolytes solutions and the thin film composite membrane chemistry has been noted in recent studies. This unique interaction is the exchange of cations across the membrane, which is present as high forward and reverse permeation of cations between the draw and feed solutions. It has been hypothesized that this behavior originates for carboxylic acid functional groups forming part of the chemical structure of the polyamide selective layer of TFC membranes. This functional group is shared by weak acid cation exchange resins which are characteristically sensitive to pH. This study seeks to further investigate the cation exchange behavior in TFC membranes with focus upon two primary goals:

1. To understand the importance of pH in affecting the bi-directional transport of cations across the selective layer of a commercial TFC FO membrane.
2. To form a TFC membrane possessing has superior resilience to cation exchange.

In this study potassium chloride has been selected as a model draw solute serving as an analog for the draw solutions envisaged for forward osmosis desalination processes. This selection eliminates the use of a draw solution which contains weak acids or bases, and the use of draw feed solutes having a metal cation allow for easy quantification of ion concentrations using atomic absorption spectroscopy. Water flux and ion fluxes were observed for a feed and draw solution kept at identical pHs, buffered at pHs of 2, 4, 6, 8, and 10. It was found that cation flux dropped significantly at pHs lower than 6.

New TFC membranes tailored for FO were prepared of differing selective layer chemistries to measure the rate of cation transport between the previously mentioned draw solution of potassium chloride and feed solution of sodium chloride. The selective layers of these membranes were formed on a composite support layer prepared from polysulfone cast over electrospun polyvinylidiene fluoride fibers. Membranes having this support layer were found to have low structural parameters and give adequate performance in FO. For the membranes prepared in this portion of the study differences in both water and ion flux were observed.

416070 Carboxyethyl Amine Sodium Salts As Novel Draw Solutes for Forward Osmosis Applications

Abstract:

For the successful implement of forward osmosis (FO) technology, development of suitable draw solutes is a critical issue, which however still faces problems of severe leakage and energy-intensive recovery. In this work, a series of novel carboxyethyl amine sodium salts (CASSs) with different carboxyl group numbers, i.e., diethylenetriamine pentapropionic acid sodium (DTPP-Na), triethylenetetramine hexapropionic acid sodium (TTHP-Na), and tetraethylenepentamine heptapropionic acid sodium (TPHP-Na), are synthesized and employed as draw solutes for FO application. Chemical structures of the synthesized CASSs are examined by ^{1}HNMR and HRMS. Their FO performances are investigated and compared in terms of different physicochemical properties. The effects of the CASS concentration on the osmotic pressure and viscosity of the draw solutions, as well as the resulted FO performance are also systematically investigated. A high water flux of 23.07 LMH and an acceptable reverse salt flux of 0.75 gMH can be achieved with 0.5 g/mL TTHP-Na draw solution under PRO mode, which is superior to most other draw solutes reported in previous literatures. TTHP-Na draw solution is further evaluated to recycle the Congo red solution to examine its applicability for waste water treatment. This study provides useful insights for the molecular design of novel high-performance draw solutes for future FO applications.

414952 Nanometric Graphene Oxide Framework Membrane: Enhancing Heavy Metal Removal Via Nanofiltration

Abstract:

GO framework membranes were developed for effective heave metal removal via nanofiltration. A thin GO framework layer was firstly constructed by the crosslinking with ethylenediamine (EDA) and subsequently assembled on a polycarbonate substrate via a pressure-assisted filtration process. The resultant membrane was further post-treated by hyperbranched polyethyleneimine (HPEI) to alter the charge property on the membrane surface. The EDA crosslinking was found not only to enhance the membrane stability, but also to enlarge the nanochannels between the laminated GO nanosheets for higher water permeability. Different post-treatment agents were compared and HPEI 60K was observed to provide a higher surface charge while lowering the mass transport resistance. In addition, all materials were prepared in aqueous solutions and the as-fabricated GO framework layer was only 70 nm thick, indicating a green and cost effective fabrication process. The newly developed membrane exhibited a high pure water permeability of 5.01 L m^{-2} h^{-1} bar^{-1} and comparably high rejection towards Mg^{2+}, Pb^{2+}, Ni^{2+}, Cd^{2+} and Zn^{2+}. These results have demonstrated the great potential of GO framework materials in waste water treatment and may provide insights for the design and fabrication of the next generation 2D-based nanofiltration membrane.

419495 Highly Permeable Dual Selective Thin Film Composite Membranes for Forward Osmosis with Viscous Draw Solutes

Abstract:

Recently, a wide range of large draw solutes have been successfully developed and employed in the forward osmosis (FO) process with the advantages of high water flux, minimum reverse flux and easy recovery. However, the high viscosity of these large draw solutes tends to suffer from more severe internal concentration polarization (ICP) effects and internal fouling of the draw solutes in the membrane sublayers. To solve this problem, a dual selective flat sheet membrane has been designed in this study in order to prevent the viscous draw solutes from entering the porous support layer and minimize the ICP effects. Dense and thin polyamide layers have been fabricated on both the top and bottom surfaces of highly porous cellulose ester support through interfacial polymerization (IP). The dual selective thin film composite (TFC) membrane demonstrated reasonably high water flux and low salt reverse flux using sodium chloride as draw solutions at the concentrations ranging from 0.25M to 1M. When tested with viscous draw solutes including sucrose, ferric citric acid complex (Fe-CA) and polyethylene glycol monolaurate (PEG 640ML), significantly mitigated ICP effects and reduced fouling propensity were observed compared to single selective TFC membranes. This study proves the great potential for the combined use of the newly developed dual selective membrane with viscous draw solutes for various FO applications.

419344 Membrane Based Separation in Metallurgy: Current Status and Perspective

Abstract:

Metallurgy is concerning metal extraction and purification. It is a very important or pillar industry in many countries worldwide. Nevertheless, in its traditional manner it has significantly negative impacts on environment. In the past decade of years, the body of literature on membrane application in metallurgy to improve its environmental friendliness and production efficiency has expanded rapidly and the new trends observed are listed below:

1) The previous ideas have been studied intensively by researchers from other countries with the advantages and disadvantages being analysed critically.

2) Incremental innovations either by modification of original technologies or transferring the technologies to the process concerning other groups of metals are emerging.

3) Various novel processes that combine membrane separation with metallurgy are proposed and investigated.

4) More and more attention is also given to tackling metallurgical waste using membrane process.

5) A distinct branch which is called "Special Industrial Membrane" mainly targeting metallurgy is growing rapidly. In addition to special polymeric membranes, inorganic and metal membranes attract great attention in recent years and are on the way of being commercialized.

With this presentation, the latest development of sustainability of metallurgy approached within the framework of membrane separation technologies will be summarized. We will goes through the above five aspects, trying to understand the strength and identify the future of membrane-based separation in this area, by describing and analyzing the real-world and potential design and applications.

419457 Nanofiltration Membrane Consisting of a Sulfonated Pentablock Copolymer Selective Layer for the Removal of Heavy Metals

Abstract:

Heavy metal pollution has become a major problem in numerous countries due to the indiscriminate discharge of untreated wastewater. The situation is expected to be further aggravated in the near future by rapid population growth and economic development. As heavy metals are toxic and non-biodegradable, their bio-accumulation in human body will cause the life dysfunction. As such, there is an urgent need to address the current situation to prevent further damage to the environment and human health. Among the methods available for wastewater treatment, nanofiltration (NF) is a feasible and promising option. However, the development of new membrane materials is constantly required for the advancement of this technology. In this study, we attempt to develop a composite NF membrane consisting of a molecularly designed pentablock copolymer rejection layer for the removal of heavy metal ions. The resultant NF membrane has a reasonably high pure water permeability (*A*) of 2.4 LMH/bar but a relatively small mean effective pore diameter of 0.50 nm, and a molecular weight cut-off of 255 Da. . The newly developed NF membrane can effectively remove heavy metal cations such as $Pb^{2}+$, Cd^{2+}, Zn^{2+} and Ni^{2+} with a rejection greater than 98.0%. In addition, the membrane also shows reasonably high rejections towards anions such as $HAsO_4^{2-}$ (99.9%) and $HCrO_4^-$ (92.3%). This performance can be attributed to (1) the pentablock copolymer's unique ability to form a continuous water transport passageway with a defined pore size and (2) the incorporation of polyethyleneimine (PEI) as a gutter layer between the selective layer and the substrate. To the best of our knowledge, this is the first reported NF membrane comprising this pentablock copolymer as the selective material. The promising preliminary results achieved in the current study may provide a useful platform for the development of new NF membranes for heavy metal removal.

435602 An Economical and UV Assisted Crumb Rubber Device for Harnessing Drinking Water from River

Abstract:

Drinking water shortage is becoming a very challenging problem for future due to the impact of calamities such as climate change, storms, burst of population and sinking water tables. In the last decade new techniques such as nano filtration, reverse osmosis, and solar distillation have been developed. Many developing countries find the new techniques expensive. In the work conducted at COMSATS Lahore, use has been made of recycled crumb rubber which is originally a source of breeding of mosquitoes. Crumb rubber fortunately acts as a very good filtration media because of its elasticity and porosity. In the experimental work, sizes of crumb rubber ranging from 1.2 – 4.2mm and 5.0 – 6.5mm were used to create columns for filtration. The average porosity was 0.50. The pilot plant consisted of a rectangular glass chamber 1.0m height and 0.62m width with a gravel support 0.3m. Glass tubes were inserted in the column to count the head loss. A UV reactor with immobilized nano-TiO_2 was attached for intake of the effluent after filtration. River water with turbidity of 238(NTU) and coliform count of 238(MPN/100ml) was used as effluent. After filtration it was found that the crumb rubber size 1.2 – 4.2mm was found to be highly effective in reducing the turbidity from 238(NTU) to 4.2(NTU) with 90% efficiency. The head loss was found to be minimum at a depth of 0.3m compared to other media depths. The turbidity decreased with increasing filtration rate of $12m^3/hm^3$. On final treatment by the photocatalytic reactor no presence of coliform was observed. The drinking water quality was in accordance with WHO specifications. The above technique has been found to be cost effective and it can be scaled up for large scale applications.

420133 Novel Processes for the Production of Olefins and Aromatics from Methane Via Methanol: A Comprehensive Process Synthesis and Global Optimization Approach

Abstract:

Recently discovered sources of shale gas, along with the technological improvements within the shale gas industry, have greatly increased the supply of natural gas in the United States. However, several challenges have emerged due to these developments, a critical one being the proper utilization of this resource into more valuable and highly demanded products via routes that are both economical and profitable. Since natural gas is the cleanest fossil fuel, these challenges present tremendous opportunities for the proper utilization of methane and have significantly increased interest in C1 chemistry across industries and within academia and government agencies.[1] To this extent, we propose novel methods for the production of high-value aromatics and olefins from natural gas via methanol. In the mid-2000s, almost all of the approximately 4.5 million metric tons per year of methanol produced in the United States utilized the synthesis gas route.

Conversion of natural gas into methanol as an intermediate for chemicals production is exceptionally promising and remains the most probable route for the widespread commercialization of natural gas based refineries in the United States. Olefins and aromatics are valuable petrochemical products and in high demand. In 2007, the worldwide consumption of ethylene was 115 million tons and the global demand of propylene was 73.5 million tons.[2] Between 2005 and 2008, the global demand for benzene, para-xylene, ortho-xylene, and meta-xylene was approximately 40, 26, 6, and 0.4 million metric tons per year, respectively.[2] Thus, opportunities exist for widespread penetration of natural-gas based chemical refineries, albeit with challenges associated with which types of chemicals to produce and which technologies to invest in.

In this work, we propose a process synthesis superstructure that contains several direct and indirect commercialized natural gas conversion technologies, including autothermal reforming and steam reforming, among others.[3,4] To address the economic barriers often associated with the utilization of alternative feedstocks, we develop a novel branch-and-bound global optimization framework that is capable of determining the optimal technologies to produce aromatics and olefins from natural gas at the highest profit.[3-5] Several commercialized and promising olefins and aromatics production, upgrading, and separation alternatives are included within this process synthesis framework.[6-9] This approach provides an adequate baseline for comparing competing technologies and identifying bottlenecks for natural gas conversion based technologies. Several case studies are presented that explore the effect of refinery capacity and product output. The key topological decisions and analysis of uncertain parameters will be discussed.

1. Mokrani T., Scurrel M. Gas conversion to liquid fuels and chemicals: the methanol route-catalysis and processes development. Catalysis Reviews 2009, 51:1-145

2. de Klerk, A. Fischer-Tropsch Refining; Wiley-VCH Verlag & Co. KgaA: Weinheim 2011

3. Onel, O.; Niziolek, A. M.; Elia, J. A.; Baliban, R. C.; Floudas, C. A. Biomass and natural gas to liquid transportation fuels and olefins (BGTL+C2_C4): Process Synthesis and Global Optimization. Industrial & Engineering Chemistry Research 2014, 54, 359-385.

4. Niziolek, A. M.; Onel, O.; Elia, J. A.; Baliban R. C.; Floudas, C. A. Coproduction of Liquid Transportation Fuels and C6_C8 Aromatics from Biomass and Natural Gas. AIChE Journal 2015, 61, 831-856.

5. Baliban, R. C.; Elia, J. A.; Misener, R.; Floudas, C. A. Global Optimization of a MINLP Process Synthesis Model for Thermochemical Based Conversion of Hybrid Coal, Biomass, and Natural Gas to Liquid Fuels. Computers and Chemical Engineering 2012, 42, 64-86.

6. Niziolek, A. M.; Onel, O.; Hasan, F. M.; Floudas, C. A. Municipal solid waste to liquid transportation fuels – Part II: Process synthesis and global optimization strategies. Computers and Chemical Engineering 2014, 74, 184-203.

7. Niziolek, A. M.; Onel, O.; Elia, J. A.; Baliban, R. C.; Xiao, X.; Floudas, C. A. Coal and Biomass to Liquid Transportation Fuels: Process Synthesis and Global Optimization Strategies. Industrial Engineering & Chemistry Research 2014, 53, 17002-17025.

8. Baliban, R. C.; Elia, J. A.; Floudas, C. A. Novel natural gas to liquids (GTL) technologies: Process synthesis and global optimization strategies. AIChE Journal 2013, 59, 505–531.

9. Baliban, R. C.; Elia, J. A.; Floudas, C. A. Biomass and Natural Gas to Liquid Transportation Fuels: Process Synthesis, Global Optimization, and Topology Analysis. Industrial & Engineering Chem Research 2013, 52, 3381–3406.

422590 Process Design and Optimization of Chemicals' Production from Biomass Feedstocks

Abstract:

Process Design and Optimization of Chemicals' Production from Biomass Feedstocks

Zhaojia Lin[1], Abhay Athaley[1], Vladimiros Nikolakis[2], Marianthi Ierapetritou[1]

[1] Department of Chemical and Biochemical Engineering, Rutgers - The State University of New Jersey

[2.] Catalysis Center for Energy Innovation & Department of Chemical & Biomolecular Engineering, University of Delaware

This work focuses on the process design and optimization of processes for the production of chemicals from lignocellulosic biomass feedstocks. The acceptance of these products in the market place depends on their competitiveness compared with petroleum-based products not only in terms of economics but also in terms of process sustainability.[6] The development of efficient and economically sustainable routes for the production of bio-based platform chemicals such as 5-hydroxymethylfurfural (HMF), furfural, levulinic acid, etc. has attracted a lot of attention.[1,2] We investigate paths based on the non-enzymatic hydrolysis of the cellulosic and hemicellulosic fractions to produce sugar mixtures followed by "one-pot" reactive extraction to convert sugar mixtures to furan derivatives. The aforementioned chemicals can be produced from glucose or xylose solutions using reactive extraction. However, the requirement for the use of organic solvents as extractants and salts that promote the partition coefficients of HMF and furfural have a negative impact on the economics and sustainability. Moreover, xylose and glucose solutions from hydrolysis steps usually contain small amounts of glucose or xylose respectively, unless further purification process is implemented.[3] Therefore, it appears more economically and environmentally attractive to simultaneously manufacture HMF and furfural from C5 and C6 mixtures that are produced from concentrated-acid biomass hydrolysis.[4,5]

Previous studies evaluated the economics and environmental impacts of these bio-products independently;[7-9] however only few studies have explored the optimization of biobased production involving both reaction and separation steps.[10] It is essential to implement flowsheet optimization to obtain the optimal operating conditions that minimizes production costs and environmental impacts both at the production but also at the separation stage. Surrogate-based optimization is a useful tool for the cases that 1) original models are expensive to evaluate; 2) the original gradient information is inaccessible for those commercial simulation tools or the noise exists so that the gradient-based algorithms cannot be accurately used. [11,12]

This work uses techno-economic analysis and life cycle assessment to design and evaluate the production alternatives. The detailed kinetics models are included in the reaction step and the separation units are simulated using Aspen Plus. At the next stage a surrogate-based optimization approach is used to achieve the optimal alternative in terms of economic and environmental criteria. More specifically, the surrogate-based models are generated to represent one or groups of original reaction and separation units; next the surrogate-based models are connected and reformulated into multi-criteria nonlinear programing problem (NLP); and finally NLP problem is solved by a commercial solver to present the pareto optimal set of solutions.

References:

1. Bozell JJ, Petersen GR. Technology development for the production of biobased products from biorefinery carbohydrates-the US Department of Energy's "Top 10" revisited. *Green Chemistry.* 2010;12(4):539-554.

2. Dutta S, De S, Saha B, Alam MI. Advances in conversion of hemicellulosic biomass to furfural and upgrading to biofuels. *Catalysis Science & Technology.* 2012;2(10):2025-2036.

3. Harris JF, Laboratory FP. *Two-stage, dilute sulfuric acid hydrolysis of wood: an investigation of fundamentals*: U.S. Dept. of Agriculture, Forest Service, Forest Products Laboratory; 1985.

4. Chheda JN, Roman-Leshkov Y, Dumesic JA. Production of 5-hydroxymethylfurfural and furfural by dehydration of biomass-derived mono- and poly-saccharides. *Green Chemistry.* 2007;9(4):342-350.

5. Zhang J, Lin L, Liu S. Efficient Production of Furan Derivatives from a Sugar Mixture by Catalytic Process. *Energy & Fuels.* 2012/07/19 2012;26(7):4560-4567.

6. Hawkins ACG, David A.; Buelter, Thomas; Wade, James; Meinhold, Peter; Peters, Matthew W.; Gruber, Patrick R.; Evanko, William A.; Aristidou, Aristos A., Assignee, Inventors; Gevo Inc. USA, assignee. METHODS FOR THE ECONOMICAL PRODUCTION OF BIOFUEL PRECURSOR THAT IS ALSO A BIOFUEL FROM BIOMASS 2009.

7. Lin Z, Ierapetritou M, Nikolakis V. Aromatics from Lignocellulosic Biomass: Economic Analysis of the Production of p-Xylene from 5-Hydroxymethylfurfural. *AIChE Journal.* 2013;59(6):2079-2087.

8. Lin Z, Nikolakis V, Ierapetritou M. Alternative Approaches for p-Xylene Production from Starch: Techno-Economic Analysis. *Industrial & Engineering Chemistry Research.* 2014/07/02 2014;53(26):10688-10699.

9. Lin Z, Nikolakis V, Ierapetritou M. Life Cycle Assessment of Biobased p-Xylene Production. *Industrial & Engineering Chemistry Research.* 2015/03/04 2015;54(8):2366-2378.

10. Torres AI, Daoutidis P, Tsapatsis M. Continuous production of 5-hydroxymethylfurfural from fructose: a design case study. *Energy & Environmental Science.* 2010;3(10):1560-1572.

11. Caballero JA, Grossmann IE. An algorithm for the use of surrogate models in modular flowsheet optimization. *AIChE Journal.* 2008;54(10):2633-2650.

12. Henao CA, Maravelias CT. Surrogate-based superstructure optimization framework. *AIChE Journal.* 2011;57(5):1216-1232.

(91c) A Review on Life Cycle Assessments of Prioritized Renewable Chemicals and Their GHG Reduction Potential

Montazeri, M., Northeastern University

Zaimes, G. G., University of Pittsburgh

Khanna, V., University of Pittsburgh

Eckelman, M. J., Northeastern University

Originally presented on: 11/9/2015 9:10:00 - 9:30:00

Abstract:

The estimated biomass production in the world is 100 petagrams (100 Gt) of carbon per year, about half in the ocean and half on land.[1] Biomass has always been a major source of energy for humankind and is presently estimated to contribute on the order of 10?14% of the world's energy supply.[2] In today's fossil fuel-driven industrial economy, production of bio-based chemicals and fuels may have numerous benefits but should also aim to decrease life cycle environmental impacts (such as GHG emissions) to be considered as favorable alternatives compared to their conventional counterparts. The RFS2 (renewable fuel standard) program administered by the USEPA regulates minimum content of renewable fuels in regular blend and sets a threshold of greenhouse gas emissions reduction for renewable fuels of least 20% for corn ethanol and a 50-60% reduction for fuels from non-corn feedstocks, cellulosic and agricultural wastes, and biodiesel.[3] While fuels have been a focus of public policy, nearly 50 million tons of bio-based chemicals are also produced annually worldwide, but there is no commensurate threshold for GHG emissions reduction of bio-based chemicals in order to define what can be considered as a renewable chemical. This presentation will review available bio-based chemical pathways and associated non-renewable energy use and life cycle GHG emissions reductions.

The US Department of Energy has conducted two separate screening analyses [4] to identify priority bio-based chemicals based on available technologies and the market demands. These chemicals can be produced from both sugar and non-sugar components of biomass resources. In this study, we conducted a thorough literature review on life cycle assessments of 11 sugar-based and 12 lignin-based building blocks reported by USDOE. 11 other bio-based chemicals that are currently in active research and development but not listed by USDOE, are also studied. This additional group can either be produced from sugar or non-sugar components of biomass. Selected studies are comprised of cultivation, preparation, intermediate and final production processes for each chemical; environmental burdens from the use and end of life phases are excluded from this study and system boundary is set to be cradle to gate. Various sources of biomass (e.g, seeds, agricultural and woody waste, pulp and paper waste streams, and algae); technology pathways (e.g., fermentation, catalytic conversion, pyrolysis, and gasification); national or regional differences; and finally life cycle assessment allocation methods (e.g., economic and mass allocation and system expansion) were considered. Tables 1a-c show the chemicals under consideration.

Table 1. a) Sugar-based and b) lignin-based building blocks from DOE reports, and c) other bio-based building blocks of interest

Sugar-based Building Blocks	Lignin-based Building Blocks	Other Building Blocks
1,4-Succinic, Fumaric and Malic Acids	Methanol/Dimethyl ether (DME)	Adipic Acid
2,5-Furan Dicarboxylic Acid	Styrenes	Polylactic Acid (PLA)
3-Hydroxy Propionic Acid	Biphenyls	Polyhydroxyalkanoate (PHA)
Aspartic Acid	Phenol	Low/ High Density Polyethylene
Glucaric Acid	Catchols, Cresols, Resorcinols	Propylene Glycol
Glutamic Acid	Vanillin	Propanediol
Itaconic Acid	Vanilic Acid	Acrylic Acid
3-Hydroxybutyrolactone	Cyclohexane	Butanol
Glycerol	Aromatic Acids	Acetic Acid
Sorbitol	Aliphatic Acids	
Xylitol/Arabinitol	Wood Preservatives	
	Formaldehyde Free Adhesives and Binders	

Life cycle GHG emission results from each of these studies are collected and compared against common petrochemical counterparts. For the comparative LCAs, GHG changes were extracted from the articles but in cases where GHG emissions of the petrochemical counterpart was not reported, Eco-Invent unit processes were used for the estimation. Other results including non-renewable energy use was also included for most of the studied scenarios.

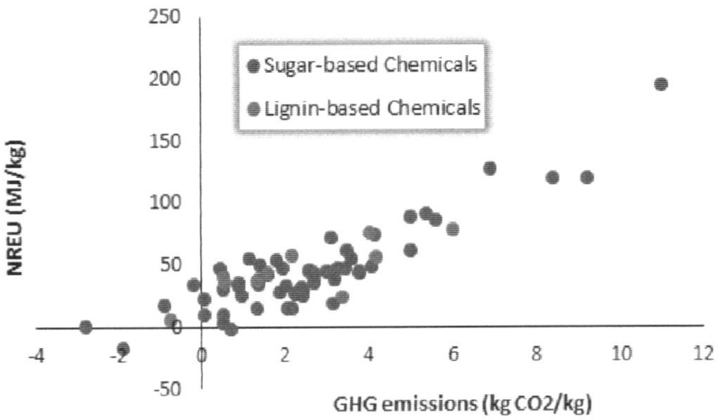

Figure 1. Non-renewable energy use versus GHG emission for sugar-based and lignin-based chemicals

Figure 1 shows the trend of non-renewable energy consumption versus greenhouse gas emissions in kg CO_2 equivalent for a functional unit of 1 kg of final product. As expected, there is a linear relationship between these parameters, as an increase in process energy consumption supplied by fossil sources will increase the emissions.

This study shows the gaps in LCA of bio-based chemicals that require more investigation, especially for lignin-based chemicals due to the complexity in primary structure of lignin and treatment methods; aspartic acid, glucaric acid, sorbitol and arabinitol are examples of sugar-based chemicals while styrene, biphenyl and cresols are examples of lignin-based chemicals included in the list but have not yet been studied on life cycle basis. Moreover, current results are dependent on processing pathways and display significant variability. The GHG results vary from >100% increase for acetic acid production from corn starch to >100% decrease for succinic acid production from corn compared to their counterparts. Even for a single chemical such as succinic acid, GHG emissions reductions range from 49% up to 100% depending on the choice of primary source, conversion and allocation methods. Bio-based chemicals that show reductions in GHG emissions also show lower fossil energy use in most cases, as shown in Figure 1.

Applying RFS2 requirements of at least 50% reduction in GHG emissions, succinic acid, PHB (polyhydroxybutyrate), xylitol, PLA (polylactic acid), polyethylene, propylene glycol and butanol show lower average values for sugar-based chemicals while methanol, PLA, PDO (propanediol), and butanol are preferable lignin-based products. Figure 2 shows these results. Average values are indicated with the full range of reported results also indicated. Blue dots show the average values while red triangles represent chemicals with no LCA studies found, indicating opportunities for further research.

Results of this study can be useful for policy makers in defining a threshold for environmental benefits should be met by renewable chemicals. There are limited data available right now to conduct a harmonization study, though. As shown in Figure 2-b, for most of the lignin-based chemicals one study was found, represented as a single point without error bars, and for the rest there are few life cycle assessment studies. On the other hand, above results are just GHG emission change, a full life cycle assessment accounting for other environmental impacts will assist finding overall benefits of renewable products. Some of the collected studies looked into other impact categories as well, but since most of the regulations are based on GHG emissions, this is the main impact category studied so far. Development of a consistent and comprehensive database for prioritized chemicals, including different pathways, sources and environmental burdens, is required for appropriate decision making.

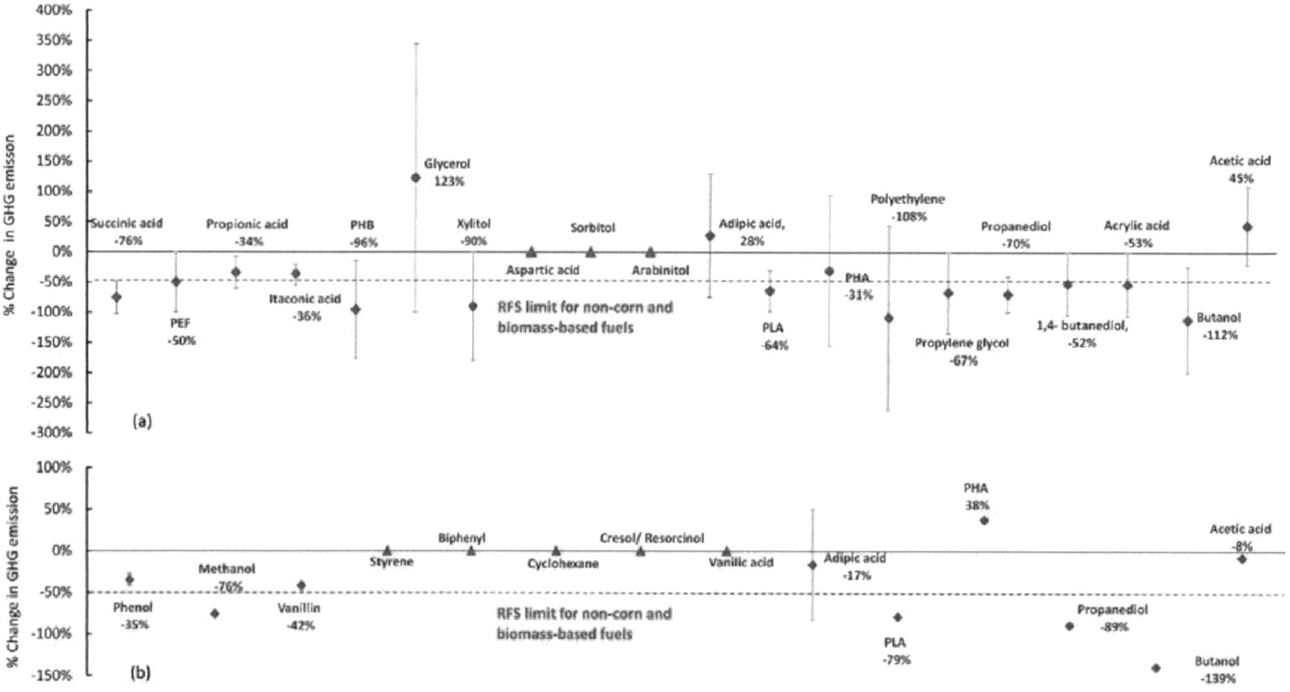

Figure 2. a) Percent change in life cycle GHG emissions of sugar-based chemicals and b) Percent change in life cycle GHG emissions of lignin-based chemicals

References:

1. Field, C. B.; Behrenfeld, M. J.; Randerson, J. T.; Falkowski, P., Primary production of the biosphere: integrating terrestrial and oceanic components. *Science* **1998,** *281* (5374), 237-240.

2. McKendry, P., Energy production from biomass (part 1): overview of biomass. *Bioresource Technology* **2002,** *83* (1), 37-46.

3. Schnepf, R.; Yacobucci, B. D. In *Renewable fuel standard (RFS): overview and issues,* CRS Report for Congress, 2010.

4. Werpy, T.; Petersen, G.; Aden, A.; Bozell, J.; Holladay, J.; White, J.; Manheim, A.; Eliot, D.; Lasure, L.; Jones, S. *Top value added chemicals from biomass. Volume 1-Results of screening for potential candidates from sugars and synthesis gas*; DTIC Document: 2004;

5. Holladay, J.; Bozell, J.; White, J.; Johnson, D., Top value-added chemicals from biomass. *Volume II?Results of Screening for Potential Candidates from Biorefinery Lignin, Report prepared by members of NREL, PNNL and University of Tennessee* **2007**.

428006 Process Simulation Models Parameterizing Multi-Sector Technoeconomic Framework to Test Viability of Emerging Bioprocesses

Abstract:

With recent advances in the development of bioprocesses, several new production schemes utilizing renewable feedstock have emerged for a wide range of chemicals. Many of these have demonstrated promising outcomes in a lab/small scale set up. Hence, lately many efforts were devoted to validate the feasibility of these processes using technoeconomic framework which intended to be used by researchers as one of the major tools to quantify the economic viability and test the potential market vulnerability.

While several exist, one of the major challenges in optimizing such a multifaceted framework is to obtain accurate process parameters, corresponding variable costs and emission estimates for the production of biochemicals. This contribution will demonstrate the manner by which these parameters are generated and subsequently linked to a Multi-Sector technoeconomic framework. Wholly, these models have the capability to capture various attributes of fundamental research and link those with the higher level economic sector, guiding the monetary investment towards the selection of robust candidate biochemicals. The resulting decision support tool can be used to study the existing market and test the viability of bioprocesses while accounting for inter and intra sector interactions in the existing value chain. The traits of this tool, bolstered with detailed process models, make it an apt venue to test various renewable process technologies, provide insights towards future production strategies and guide the research to produce bio-based products.

434302 Biolubricant: Lignin in Choline/Amino Acid Ionic Liquids

Abstract:

Lignocellulosic biomass is the most abundantly available raw resource on the earth for the production of valuable products. Utilization of cellulose in versatile promising applications has been intensely investigated in last a few decades. However, the utilization of lignin is greatly restricted due to the its resistant to degradation, being held together with strong chemical bonds and also appears to have a lot of internal hydrogen bonds. Lignin fractionation from biomass has been demonstrated in choline based ionic liquids(ILs), however, the high cost of the ILs necessitates its recycle in industrial process which is difficult due to the large energy consumption. In this work, we present a new application for lignin dissolved in Choline based amino acid ILs as a green bio-lubricant. The addition of lignin could improve choline based ILs' anti-wear properties and friction stability. Various characterization techniques including differential scanning calorimetry, thermogravimetric analysis, and rheology are used to study the effect of lignin on the thermal, physical and tribological properties of lignin/choline based ILs. The results show that the lignin could enhance the physiochemical properties simultaneously.

437859 Water Desalination By Shock Electrodialysis

Abstract:

The development of cost and energy efficient water purification systems, specifically desalination systems, is one of the most critical engineering challenges today. Recently, desalination has been used to solve the drought crisis in California, but it is costly and energy intensive [1]. Shock electrodialysis is a newly developed method that uses ion concentration polarization (ICP) zones and deionization shocks in porous media near an ion selective membrane to purify water. In this project, we demonstrate experimentally the feasibility of shock electrodialysis in water desalination. An overlimiting current, a current that is more than the diffusion limited current, is applied through a weakly charged porous medium (a silica glass frit with pore sizes ranging from 0.75 to 1 micron) sandwiched between two cation-selective membranes (Nafion), producing a sharp concentration gradient or "shock"[2]. The cations move through the cation-selective membrane toward the cathode while the anions get trapped on the cation-selective membrane on the anode side. This movement of ions produces a sharp concentration gradient with high ion concentration near the anode side and low ion concentration near the cathode side. Flowing saltwater through the porous medium and adding a splitter that splits the flow into two streams at the end can separate the high ion concentration (brine) stream and the low ion concentration (fresh) stream. This separation allows for continuous desalination. Using this device, we were able to remove up to 99+% of ions from NaCl feed solutions of various concentrations (1 mM, 10mM, and 100mM) at a flowrate of 75 uL/min. Keeping the flow rate constant at 75 uL/min and varying the concentrations revealed that plots of the percentage of ion removal versus the applied current divided by the limiting current for all three concentrations overlapped, agreeing with the theory. By varying the flow rate and keeping the concentration of NaCl constant at 10mM, it was seen that at lower flowrates, there was more ion removal at the same applied current, agreeing with the theory because a lower flowrate allows the solution to spend more time in the device which allows for more ion removal. The 37.5 uL/min and 75 uL/min flowrates achieved 99+% ion removal at the highest applied current (4080 uA) while the 150 uL/min flowrate achieved 90% ion removal. Overall, preliminary results show good desalination factors, reasonable efficiency, and unique separation capabilities. These unique separation capabilities stem from the feed water flowing through the glass silica and the large electric field applied. The small pore size in the glass silica frit is able to separate out large particles and the large electric field has the potential for disinfection; both of these qualities may find applications in compact water treatment and brine concentration.

[1] Roth, S.; California's last resort: Drink the Pacific; Desert Sun; 2015. http://www.desertsun.com/story/news/environment/2015/04/20/californias-last-resort-drink-pacific/26081355/

[2] Mani, A.; Zangle, T.; Santiago, J.G. On the Propagation of Concentration Polarization from Microchannel-Nanochannel Interfaces Part I: Analytical Model and Characteristic Analysis. Langmuir 2009, 25 (6), 3898-3908

437635 Exergy Analysis of a Power Plant in Abu Dhabi (UAE)

Abstract:

The first objective of this research was to conduct an exergy analysis for a power-generation plant in Abu Dhabi (UAE) in order to determine the main source of its irreversibilities. Our results indicated that the combustion chamber was the main factor (70.2%) of the total exergy destruction of the plant. On the other hand, the compressor had the lowest contribution towards exergy destruction (12.4%). In the second part of this investigation, Aspen Hysys with the Soave-Redlich-Kwong (SRK) equation of state were utilized in order to simulate the effects of high temperatures and absolute humidity of ambient air on the exergy efficiency of the different equipment in the power plant. Our data showed that the temperature of ambient air had more negative effects on the performance of the plant than its absolute humidity. The exergy efficiency of the combustor and compressor decreased with temperature while the exergy efficiency of the turbine increased with temperature. Finally, a comparative study on performance of the power plant between design conditions (T=288K, RH=60%) and summer conditions (T=316K, RH=50%) in Abu Dhabi indicated that the power plant lost 4.66 % of its net power output and 4.61 % of its exergy efficiency.

438525 Insights into the Hydrothermal Stability of ZSM-5 Under Relevant Biomass Conversion Reaction Conditions

Abstract:

Zeolite catalysts used for the conversion of carbohydrates to renewable platform chemicals (e.g., glucose to HMF) in the condensed phase are shown to be sensitive to the presence of inorganic salts which alter the zeolite surface chemistry. Although zeolites such as ZSM-5 have been shown to be stable in de-ionized water, the presence of NaCl in the reaction mixture at dilute and saturated levels enhances the hydrolysis of the Si-O-Al bridges and the release of Al3+ species. These dissolved Al3+ species catalyze the conversion of glucose through homogeneous catalytic processes and obscure the apparent reactivity of the zeolite catalyst. Comparative XRD, solution-state NMR, and ICP-MS show that the reactive aluminum species in solution are leeched in a time-limited process from Si-O-Al bonds involved in extra-framework species and/or at defect sites without affecting the zeolite lattice

EFFECTS OF SEASON AND HEATING MODE ON IGNITION AND BURNING BEHAVIOR OF TEN SPECIES OF LIVE FUEL MEASURED IN A FLAT-FLAME BURNER SYSTEM

Samantha A. Smith, Jonathan R. Gallacher, and Thomas H. Fletcher

Abstract

The effects of season and heating mode on ignition and burning behavior of living vegetation were studied in a flat-flame burner system with a radiant panel. The goal was to identify what plant characteristics have the greatest influence on burning behavior and to understand the effects of heating mode on ignition and burning. Experiments were performed for ten species over a two-year period, with 25 runs completed each month for each species. A flat flame burner (1000°C, 10 mol% O_2) and radiant panel (50 kW/m^2) provided the convection and radiation sources, respectively. Time-dependent mass, surface temperature and flame characteristics were measured. The effect of moisture content on ignition time varied with species. Ignition time showed a strong dependence on heating mode, with broadleaf species showing a much stronger response to added radiation than non-broadleaf species. No samples exposed to the radiant panel alone ignited. Heating mode and moisture content affected flame characteristics and the relationship depended upon season and species. The observed relationship between ignition time and moisture content did not agree with the relationship observed in wet wood. Moisture content alone did not adequately describe seasonal changes in ignition behavior. However, by including other factors in the prediction model (e.g. surface area and mass), it is possible to predict live-fuel ignition. This implies that plant function plays a role in the burning behavior of live fuels.

1. Introduction

Ignition of wood and other cellulosic fuels has been studied for over 100 years. Research has largely focused on requirements for ignition, conditions during burning, and predictive modeling techniques (including rate of spread calculations). The ultimate goal in wildland fire research is two-fold: (1) to understand the physical phenomena that occur within wildland fires, and (2) to develop models that can predict wildland fire behavior. This study presents some of the results from a two-year project to study ignition and burning phenomena relative to seasonal changes in plant growth and the mode of heat transfer used for fuel heating.

Ignition can occur when a fuel sample is heated to the point where pyrolysis rates are high enough to support a gaseous flame *and* a gas phase flammable mixture exists. Researchers often measure ignition temperature and time. These values are then used as empirical estimates of the surface temperature at which pyrolysis rates can support a continuous flame and the time required to reach that temperature, respectively. The bulk of the work has focused on ignition temperature. Experimental conclusions to date are mixed. Babrauskas [1, 2] compiled the results of ignition temperature experiments on wood fuels and foliage, respectively. After eliminating the experiments in which the fuel sample was pressed against a hot surface, the reported ignition temperatures ranged from 200-530°C for wood and 201-450°C for foliage. Babrauskas noted the large amount of scatter in the data and suggested that, in addition to variations in experimental setup and measurement techniques, sample condition (e.g. moisture content and size) and species could affect ignition temperature.

Wildland fire observations that species burn differently support Babrauskas's postulate that plant species could be one source of variation in measured ignition temperatures [3]. However, results by Susott [4] showed that material ground from various plant species has the same heat of combustion and similar TGA (thermogravimetric analysis) pyrolysis mass release curves, and should therefore burn similarly. Thus, one possible explanation for the observed differences in ignition properties is the shape and structure of the plant and the effect shape has on heat and mass transfer. However, this explanation has not been tested experimentally. Most empirical correlations used to predict ignition behavior, particularly for live fuels, are species specific [5-8]. Work must be done to understand the differences in ignition behavior between various species.

Investigation of the effect of moisture content on ignition has been studied extensively and supports Babrauskas's postulate that sample condition affects ignition. Most of the work has focused on dead or woody fuels, with the result that moisture increases both ignition time and temperature [5, 6, 9-11]. The results for live fuels are fewer and less consistent [3, 12-15]. Many fire spread models use moisture content as a predictor for ignition. These models have varied success with dead fuels but little success with live fuels [13]. One possible reason for the difference in ignition between live and dead fuels and the performance of fire spread models in predicting fire behavior in live fuels is the difference in water storage between live and dead plants and the resulting treatment of water storage in fire spread models. Water evaporation in dead fuels has been assumed complete in fine fuels once the sample temperature passes 100°C [16, 17], but Fletcher et al. [3] showed there is still a significant amount of moisture in live fuels when ignition occurs. Pickett [18] showed water release still occurring at surface temperatures in excess of 200°C and Prince [14] showed significant differences in the temperature profiles of live and dead foliage during ignition and burning even with the same moisture content. Work by McAllister et al. [13] showed significant differences in the ignition behavior between live and dead pine needles. These differences have led researchers to postulate that there is significant interaction between the free water and the cells in live plants that does not occur in dead plants [13, 19]. Finney et al. [20] postulated that water release in live fuels is not complete until the breakdown of the cellular structure occurs. Still other work has been done indicating root structure [21], plant dry mass [22], and chemical composition [13, 23] could have a larger effect on ignition of live fuels than moisture content, though results are mixed in work to quantify the effect of chemical composition (e.g. [15]). Thus, a fundamental understanding of the physical processes that drive live fuel combustion is both absent and necessary if predictive models are to be developed.

Another of Babrauskas's postulates was that differences in ignition temperature arise due to differences in experimental setup. This is particularly true when looking at the different heat transfer mechanisms. Much of the work looking at heat transfer mode has focused on dead and woody fuels [10, 17, 24-31]. Experiments performed by Rothermel [17] showed fuel pre-heating in no-wind and backing-fire situations, leading to the conclusion that radiation drives fire spread. This conclusion and his resulting fire spread model established a precedent that radiation is the dominant form of heat transfer in wildland fires. However, other experiments have shown that significant amounts of pyrolyzates are not formed at the fuel temperatures associated with radiant pre-heating [32]. Anderson [33] concluded that radiant heat flux can provide no more than 40% of the energy required for sustained fire spread. Still other work has shown flame propagation to depend strongly on direct flame contact with un-burned fuel [34, 35]. Many researchers have concluded that radiation heat transfer dominates in large fires [24, 31, 36, 37] and fires with little to no wind in homogeneous fuel beds [28, 38-40], but the relative effect of radiation and convection for fires outside these conditions is still unknown [28, 40]. Only a limited amount of work has been performed for live fuels and foliage [13, 41]. Work to quantify the contributions of radiation and convection in live-shrub combustion is necessary to understand the basic theory of fire spread and to develop a model that accounts for both mechanisms of heat transfer.

The aim of this project is to explore the effect of season and heating mode on ignition and burning behavior to better understand what physical processes drive fire spread in live shrubs.

2. Methods

Combustion experiments were performed on ten species every month for two years. Each month 25 replicates were performed for each species. The ten species consisted of three species from California, three from Florida, two from Utah, and two from Montana. The plants from California consisted of chamise (*Adenostoma fasciculatum*), Manzanita (*Arctostaphylos glandulosa*), and Ceanothus (*Ceanothus crassifolius*). Gallberry (*Ilex glabra*), fetterbush (*Lyonia lucida*), and sand pine (*Pinus clausa*) originated from Florida, while big sagebrush (*Artemisia tridentata*), and Gambel oak (*Quercus gambelii*) are from Utah. Finally, lodgepole pine (*Pinus contorta*) and Douglas-fir (*Pseudotsuga menziesii* var. *glauca*) are from Montana. Since most of the samples had to be collected from other locations, they were placed in a sealed plastic bag and shipped overnight, where they were placed away from light with the bag still closed until that species could be tested.

Experiments were performed using a water cooled, flat-flame burner (FFB) and radiant panel (see Figure 1). The FFB produced post-combustion gases at 1000 °C and 10 mol% O_2 while the radiant panel provided 50 kW m^{-2} at the sample location. A glass cage prevented ambient air entrainment. In order to perform the experiments, the sample was suspended within the cage using an alligator clip and holding rod. This rod was placed on a Mettler Toledo XS204 Cantilever mass balance, which recorded any differences in mass as the specimen burned or experienced pyrolysis. Each month, experiments were performed using three heating cases, namely convection, radiation, and combined convection plus radiation. In all cases there was a visual camera placed in front of the sample and an infrared camera placed above the sample (see Figure 1).

Transient flame, mass, and surface temperature data were collected during each run. Flame characteristics were extracted from these data, including maximum flame height (cm), time to maximum flame height (s), ignition temperature (°C), maximum ignition temperature (°C), ignition time (s), and burn out time (s). There were also many measurements of the specimen itself, such as the density, surface area, moisture content, length, width, thickness, etc.

Both single and multiple parameter prediction models were developed to describe each of the flame characteristics listed above, namely time to ignition (t_{ig}), surface temperature at ignition (T_{ig}), maximum surface temperature at ignition ($T_{ig,max}$), maximum flame height (MFH), time to maximum flame height (t_{MFH}) and burnout time (t_{BO}). Ignition has been shown to be a local phenomenon [19]. Surface temperature has also been shown to have large spatial variations during the burning of live foliage [19]. However, many ignition temperature measurements do not measure the spatial variation in surface temperature. Correlations for both average surface temperature at ignition and maximum surface temperature at ignition (assuming the location of maximum temperature corresponds to the localized ignition point) were developed. One-parameter models were developed using simple linear regression. The F-statistic, p-value, and confidence interval on the slope were used to determine if the slope term and regression model were significant.

Forward and backward stepwise regression was used to develop correlations with multiple parameters. To do this, 500 correlations were developed by randomly selecting four parameters for use in each model and by randomly selecting a transformation of the data for those parameters. To develop the models, the parameters were first assigned to groups based on type of measurement to reduce the chance of multicollinearity. The groups were: (1) moisture content, relative moisture content and water mass; (2) fresh mass and dry mass; (3) length, width and surface area; (4) needle length, stem diameter, thickness and density; and (5) lipid content, volatile fraction, fixed carbon and ash content. For each correlation, four of the five groups were randomly chosen and then a parameter from each of those four groups was randomly selected. Once the parameters were identified, a data transformation was applied to the data for each parameter. The transformations were: (1) no change, (2) square root of the data, (3) square of the data, (4) natural log of the data, and (5) inverse of the data. Once the data and data transformations were assigned, stepwise regression was performed to develop a correlation to fit the selected data to each of burning characteristics listed above. Once the 500 models were created, the model with the highest adjusted R^2 for each burning characteristic for each species was identified.

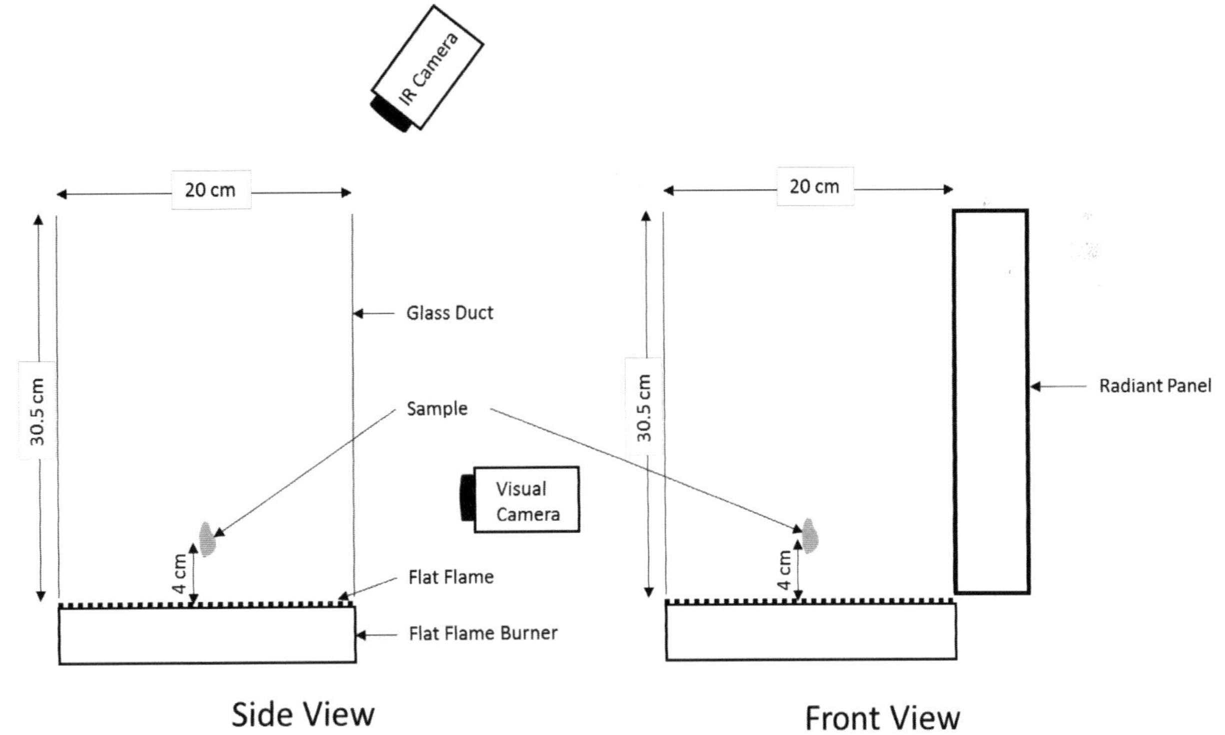

Figure 1. Experimental apparatus: flat flame burner, radiant panel, IR camera, video camera, glass cage to prevent ambient air entrainment, sample location, mass balance and sample holding rod. Post-flame gases are 1000 °C and 10 mol% O_2.

3. Results and Discussion

It was predicted that an increase in the moisture content of a species would result in a longer ignition time and would closely resemble the burning of wet wood. This was not the case. Table 1 illustrates how live fuels responded to different burning modes during the year for different flame characteristics. If there was a statistically significant positive or negative slope at the 95% confidence level, then a P or N was indicated, respectively. Zeros indicate that there was no statistically significant relationship. Overall, the results from the live fuels were different from the expected behavior with wet wood illustrating that one-parameter models do not work for live fuels. Ceanothus, Gambel Oak, Douglas-fir, and Chamise, however, did exhibit a positive correlation to the wet wood behavior at the 95% confidence level.

Table 1. Significance of yearly trends by species

	Ignition Time (s) vs MC		**MFH (cm) vs MC**		**Ignition Devol (%/s) vs MC**		**Ignition Devol (%/s) vs Ignition Time (s)**	
Species	Conv*	Comb**	Conv	Comb	Conv	Comb	Conv	Comb
Manzanita	0	0	0	0	0	0	N	0
Ceanothus	P	P	0	0	0	0	0	0
Douglas-fir	P	0	N	0	0	0	0	0
Gambel Oak	P	P	N	N	0	0	0	0
Fetterbush	0	0	0	0	0	0	N	0
Gallberry	0	0	0	0	0	0	N	0
Sand Pine	0	P	0	0	0	0	0	0
Chamise	P	P	0	0	0	0	0	0
Sagebrush	0	0	0	0	0	0	0	0
Lodgepole Pine	0	0	0	0	0	0	0	0
Wet Wood (expected)	P	P	N	N	P	P	P	P

*Conv means convection-only experiments; **Comb means combined convection and radiation

The models created using stepwise regression were much better than the one-parameter models. Table 2 shows the adjusted R^2 values for the flame characteristics measured with these new models. The average adjusted R^2 value for all the models is 0.406. However, the models for temperature account for much less of the data variation than the models for the other flame characteristics; the average adjusted R^2 value for the non-temperature models in 0.569. This is partly due to the fact that samples bend and move during burning and thus the entire sample surface was not visible to the IR camera during the run.

Table 2: Adjusted R^2 values when regressing flame characteristics using stepwise models.

Species	t_{ig}	T_{ig}	$T_{ig,max}$	MFH	t_{MFH}	t_{BO}
Manzanita	0.80	0.527	0.568	0.740	0.919	0.747
Ceanothus	0.794	0.481	0.372	0.743	0.905	0.676
Douglas-fir	0.303	0.571	0.692	0.342	0.254	0.50
Gambel Oak	0.664	0.472	0.186	0.432	0.86	0.81
Fetterbush	0.537	0.379	0.058	0.431	0.417	0.459
Gallberry	0.733	0.152	0.223	0.683	0.743	0.511
Sand Pine	0.434	0.294	0.247	0.58	0.624	0.639
Chamise	0.554	0.286	0.305	0.617	0.502	0.449
Sagebrush	0.449	0.285	0.247	0.403	0.589	0.403
Lodgepole Pine	0.489	0.315	0.309	0.384	0.36	0.30

Figure 2 illustrates how ignition time and moisture content vary throughout the year for Douglas-fir and Manzanita with different heating modes (convection-only, and combined convection and radiation). It is observed that there is no statistically significant relationship between moisture content and ignition time. Manzanita had approximately the same moisture content in July and November, but the ignition times in July were about half as long as the ignition times in November. Douglas-fir moisture content doubled from May to September, yet there was no significant change in the ignition time. These examples point toward the initial observation that the single parameter model of just moisture content cannot provide an accurate prediction for live fuels (see Table 1). Since the moisture content isn't the only thing that is impacting the burning of the live fuels, it is logical to consider the plant's physiological structure and how that can change the way it burns. The structure of the plant does change throughout the seasons and that can change how the species stores the water and how the leaves are shaped. These changes can help account for the changes that are not accounted for by just the water in the plant. It is important to note that, while a study of how physiological factors impact the burning of the live fuels is needed to develop a deeper theoretical understanding of live fuel combustion, it is still possible to predict burning behavior with statistical models that use the same inputs as current fire spread models.

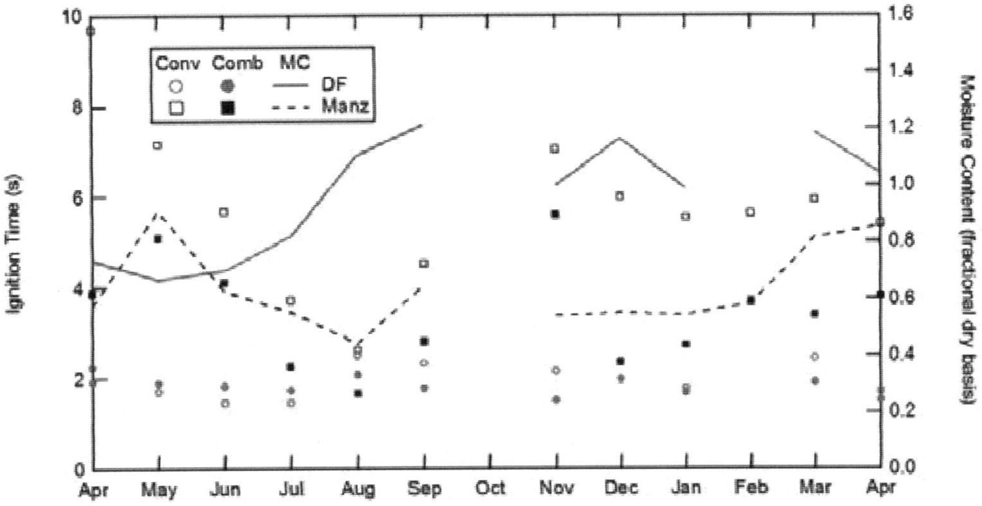

Figure 2. Ignition time and moisture content versus month for manzanita and Douglas-fir. Closed symbols indicate radiation and convection ignition while open symbols indicate convection-only ignition.

In addition to the models that were produced, there was data collected on the different heating modes and how they impact the burning of the live fuels. Table 3 shows the percentage of months there was a statistically significant difference between the convection-only and convection and radiation runs for each species. There is a noticeable difference between the broadleaf species and the needle species for the ignition time and time to maximum flame height due to the added radiation. Radiation alone was never enough to ignite fuel samples without a pilot ignition source. The heat did, however, cause the samples to undergo pyrolysis and then char oxidation.

Table 3. Effect of heating mode on ignition variables. Table entries indicate the percentage of months that radiation and convection ignition differed significantly convection-only ignition at a 95% confidence level

Species*	Ignition Time	Time to Max Flame Height	Maximum Flame Height	Fraction Left at Ignition	Fraction Loss at Ignition
Manzanita	92	83	83	25	42
Ceanothus	100	91	100	45	18
Gambel oak	83	100	83	50	17
Fetterbush	100	100	8	17	25
Gallberry	92	100	0	17	58
Dougals-fir	9	27	27	0	9
Sand pine	25	25	8	0	17
Chamise	33	25	17	17	0
Sagebrush	22	22	22	11	11
Lodgepole pine	50	25	13	38	13

*The first five species listed are broadleaf species, the other five are non-broadleaf.

4. Summary and Conclusions

Ignition and burning behavior for ten live fuels were studied in a flat-flame burner apparatus. Experiments were performed over a two-year period to test the effect of season (specifically moisture content) on ignition and burning behavior. The effect of heat transfer mode was also tested by performing some of the seasonal experiments with a convection-only heat source, some with a radiation-only heat source, and some with both heat sources.

Results comparing ignition and burning characteristics with moisture content and season were mixed. Ceanothus, Gambel oak, Douglas-fir and chamise all exhibited a positive correlation between ignition time and moisture content while the other six species show no correlation at a 95% confidence level, indicating a simple relationship between moisture content and ignition is not adequate to describe ignition in live fuels. However, linear stepwise models capture much of the variability in ignition behavior. The ignition behavior of live fuels in different seasons but at the same moisture content was different. These results suggest a significant difference between live fuels and dead fuels that isn't explained by moisture content. An in-depth study involving careful combustion experiments and detailed physiological measurements is necessary to develop a theoretical understanding of fire spread in live fuels. In the absence of a theoretical understanding, simple statistical models can still be developed that describe fire behavior accurately and that use as inputs the same information currently used in most fire models.

Heating mode results were clearer. Ignition occurred in none of the unpiloted radiation-only experiments. These results do not indicate radiation was unimportant, but rather that additional modes of heat transfer were needed. Inclusion of a radiant flux in the convective environment of a flat-flame burner significantly decreased observed ignition times for broadleaf species but not for non-broadleaf species. Maximum flame height time showed the same dependence as ignition time for all species while maximum flame height showed the opposite dependence, except for fetterbush and gallberry which showed almost no difference between convection-only and radiation with convection experimental runs. Mass loss rate and fraction remaining at ignition were affected much less by the differences in heating than other ignition characteristics.

5. References

[1] Babrauskas, V., J. of Fire Protection Eng., 12 (2002) 163-189.
[2] Babrauskas, V., Ignition handbook, Fire Science Publishers, 2003.
[3] Fletcher, T. H., Pickett, B. M., Smith, S. G., Spittle, G. S., Woodhouse, M. M., Haake, E.Weise, D. R., Comb. Sci. and Tech., 179 (2007) 1183-1203.
[4] Susott, R. A., For. Sci., 28 (1982) 404-420.
[5] Xanthopoulos, G.Wakimoto, R. H., Canadian J. of Forest Research, 23 (1993).
[6] Dimitrakopoulos, A. P.Papaioannou, K. K., Fire Tech., 37 (2001) 143-152.
[7] Smith, S. G., Effects of moisture on combustion characteristics of live California chaparral and Utah foliage, Brigham Young University (2005).
[8] Shen, C., Application of fuel element combustion properties to a semi-empirical flame propagation model for live wildland Utah shrubs, Brigham Young University (2013).
[9] Gill, A. M.Moore, P. H. R., Ignitability of leaves of Australian plants, A Contract Report to the Australian Flora Foundation, Centre for Plant Biodiversity Research, 1996.
[10] Moghtaderi, B., Novozhilov, V., Fletcher, D. F.Kent, J. H., Fire Safety J., 29 (1997) 41-59.
[11] Catchpole, W. R., Catchpole, E. A., Tate, A. G., Butler, B. W.Rothermel, R. C., A model for the steady spread of fire through a homogeneous fuel bed, 2002.
[12] Engstrom, J. D., Butler, J. K., Smith, S. G., Baxter, L. L., Fletcher, T. H.Weise, D. R., Comb. Sci. and Tech., 176 (2004) 1577-1591.
[13] McAllister, S., Grenfell, I., Hadlow, A., Jolly, W. M., Finney, M.Cohen, J., Fire Safety J., 51 (2012) 133-142.
[14] Prince, D., Measurement and modeling of fire behavior in leaves and sparse shrubs, Brigham Young University (2014).
[15] Alessio, G. A., Penuelas, J., De Lillis, M.Llusia, J., Plant Biology, 10 (2008) 123-128.
[16] Albini, F. A., Proc. Comb. Inst., 11 (1967) 553-560.
[17] Rothermel, R. C., A mathematical model for predicting fire spread in wildland fuels, Research Paper INT-115, USDA Forest Service, 1972.
[18] Pickett, B. M., Effects of moisture on combustion of live wildland forest fuels, Brigham Young University (2008).

[19] Prince, D.Fletcher, T. H., A combined experimental and theoretical study of the combustion of live vs. dead leaves, Combustion Institute - Western States Section (WSS/CI), 2013.
[20] Finney, M. A., Cohen, J. D., McAllister, S. S.Jolly, W. M., Int. J. of Wildland Fire, 22 (2013) 25.
[21] Pellizzaro, G., Duce, P., Ventura, A.Zara, P., Int. J. of Wildland Fire, 16 (2007) 633-641.
[22] Jolly, W. M., Hadlow, A. M.Huguet, K., Int. J. of Wildland Fire, 23 (2014) 480-489.
[23] Pyne, S. J., Andrews, P. L.Laven, R. D., Introduction to Wildland Fire, 2nd ed, Wiley, 1996.
[24] Simms, L., Comb. Flame, 4 (1960) 293-300.
[25] McCarter, R. J.Broido, A., Pyro., 2 (1965) 65-85.
[26] Simms, L.Law, M., Comb. Flame, 11 (1967) 377-388.
[27] Pagni, J., J. of Heat Trans., 97 (1975) 153-155.
[28] Morandini, F., Santoni, P. A.Balbi, J. H., Fire Safety J., 36 (2001) 519-543.
[29] Dupuy, J. L., Maréchal, J.Morvan, D., Comb. Flame, 135 (2003) 65-76.
[30] Reszka, P.Torero, J. L., Exp. Therm. and Fluid Sci., 32 (2008) 1405-1411.
[31] Silvani, X.Morandini, F., Fire Safety J., 44 (2009) 279-285.
[32] Cohen, J. D.Finney, M. A., VI International Conference on Forest Fire Research, (2010).
[33] Anderson, H. E., Heat transfer and fire spread, in Research Paper INT-69. 1969, USDA Forest Service.
[34] Vogel, M.Williams, F. R., Comb. Sci. and Tech., 1 (1970) 429-436.
[35] Carrier, G. F., Fendell, F. E.Wolff, M. F., Comb. Sci. and Tech., 75 (1991) 31-51.
[36] Balbi, J.-H., Rossi, J.-L., Marcelli, T.Santoni, P.-A., Comb. Sci. and Tech., 179 (2007) 2511-2537.
[37] Paudel, A., Application of one dimensional turbulence (odt) to model fire spread through biomass fuel bed, Brigham Young University (2013).
[38] Morvan, D.Dupuy, J. L., Comb. Flame, 127 (2001) 1981-1994.
[39] Morvan, D.Dupuy, J. L., Comb. Flame, 138 (2004) 199-210.
[40] Sullivan, A. L., Ellis, P. F.Knight, I. K., Int. J. of Wildland Fire, 12 (2003) 101-110.
[41] Stocks, B. J., Alexander, M. E., Wotton, B. M., Stefner, C. N., Flannigan, M. D., Taylor, S. W., Lavoie, N., Mason, J. A., Hartley, G. R., Maffey, M. E., Dalrymple, G. N., Blake, T. W., Cruz, M. G.Lanoville, R. A., Canadian J. of Forest Research, 34 (2004) 1548-1560.
[42] ASTMStandardD4442-07, Test methods for direct moisture content of wood and wood-based materials, in ASTM International. 2007: West Conshohocken, PA.
[43] Weber, R. O., Prog. Energy Combust. Sci., 17 (1991) 67-82.
[44] Sullivan, A. L., Int. J. of Wildland Fire, 18 (2009) 349-368.

438326 Characterization and Particle Sizing of the Composition of E-Cigarette Aerosol

Abstract:

In recent years, electronic cigarettes, or e-cigarettes, have rapidly gained immense popularity throughout the world and specifically the United States. Despite the increase in use, little is known about the health effects of these vaporizers, compared to traditional cigarettes. The e-cigarette uses a wick and heated coil to atomize e-liquid, creating the aerosol inhaled by the user. E-liquids are composed of propylene glycol, glycerin, usually nicotine, and various flavoring agents. E-cigarettes were customized in order to simulate a puff on the device by pushing a known volume of air through. An Automated Puffer Machine (APM) was designed to control power delivered to the coil, puff volume, puff frequency, and puff duration. An Electrostatic Classifier (EC) and Condensation Particle Counter (CPC) then sampled a diluted puff, analyzing particle size and concentration. A Poly Urethane Foam (PUF) sampler was used to separate and trap the chemical components of the e-cigarette emission. The PUF was then placed in a volatile solvent to dissolve the particulate and vapor components, and the solvent was analyzed using gas chromatography with mass spectroscopy. The PUF sample was used because is closely emulates what a user would normally inhale.

Work has also been done to quantify and describe the aerosol released by electronic cigarettes in order to create profiles of the particles released into the atmosphere. Examples include work to quantify particle size distribution and chemical content of the aerosol in large gas chambers and looking at particle size and distribution with rapid sampling. Particle size of the aerosol was a bimodal distribution, with main peak averages of 30 nanometers and 80 nanometers. Research isolated the components of the e-liquid attempting to sort which vaporized component contributed to the collected particle size profile. Further research attempted to determine the amount of nicotine available within both the vapor and particle components of the aerosol and identify the exact composition of the aerosol created by an e-cigarette when the e-liquid is vaporized.

This study has found that a variety of e-liquids ranging in nicotine contents and flavors consistently produce a bimodal distribution of aerosol particle sizes. Furthermore, this study concludes that the proportion of components effects the concentration of particles of the aerosol.

438306 Ammonia Removal from Aquaculture Stocking Water

Abstract:

A continuous membrane system capable of maintaining low levels of total ammonia nitrogen (TAN, 3, and 4 +) in aquaculture holding and transport vessels is being developed. Ammonia is the major metabolite of fish, therefore, adequate TAN removal is necessary to maintain the health and productivity of aquatic organisms. High levels of ammonia result in the damage of gas exchange tissues, leading to increased mortality and decreased average harvest weight. Membrane systems are desirable for their scalability, lack of compaction and channeling, as in packed beds, and their ability to stabilize additives used during functionalization. Zeolite 13X demonstrates efficient 4 + exchange with + in aqueous solutions. Zeolite 13X is inexpensive and functions well as an ion exchange material with high ammonium exchange capacities. In order to improve the stability of zeolite 13X in water and develop a continuous TAN removal scheme, mixed-matrix membranes have been fabricated in the presence and absence of porogens. Porogens have been used to investigate the effect of pore size and pore density on the membrane structure and the membrane performance. The membranes with porogens exhibited higher flux but lower ammonia removal due to decreased residence time. Porogen-free membranes had a higher ammonia removal capacity due to longer residence time. The ammonia removal capacity was found to correlate well with the amount of zeolite 13X incorporated. An optimal formulation will maximize flux and available zeolite surface area, minimize residence time, and retain large ammonia loadings after a large volume of liquid has been treated.

438394 Impact of Chemical Dopants and Passivation Schemes on Carbon Nanotube Sheet Conductivity

Abstract:

Carbon nanotubes (CNTs) are being investigated as an alternative to conventional metallic conductors. Improving the electrical performance of bulk CNT materials through the use of chemical dopants is an active area of research. The present work focuses on doping efficacy and dopant stability under various environmental conditions. CNT sheet material was doped with a series of compounds of varying structure, and a 3 to 8-fold increase in electrical conductivity was observed after doping with potassium tetrabromoaurate, iodine, or iodine monobromide. The general trend for the doped samples in ambient conditions was a steady decrease in electrical conductivity over several months. These experiments served as an initial assessment of doped CNT sheet stability for conventional conductor applications and demonstrate the necessity towards passivation schemes for this emerging technology. Different encapsulation strategies are currently being explored to retain the conductivity of the films when doped. The conductivity of these samples is being systematically measured while storing them under ambient conditions. Consideration for the encapsulant type, deposition technique, and solvent effects (which may dissolve the dopants from the CNT network) will be characterized. A summary detailing the progress to date will provide an overview of both the best methods to enhance electrical conductivity in CNT sheets as well as the suitable passivation schemes for practical operation.

438530 Îœfilms: Dynamics of Thin-Films Under Physiological Fluids and Shear Flow

Abstract:

Thin-film drug-delivery is an effective alternative to conventional tablets, capsules and liquids as it improves medicinal bioavailability, efficacy, dosage accuracy and onset of effectiveness. Typically, it consists of a thin, dissolving film that orally administers drugs. The significantly large surface area of the thin-film causes rapid disintegration, releasing drugs into systemic circulation via absorption through the oral mucosa. Thin-films are currently studied under static conditions that do not account for the influence of dynamic variables like shear stress due to blood flow, changes in electric potential and interactions with biological fluids on drug delivery. Here we seek to study films under dynamic physiological conditions that closely mimic those within the body, using a microfluidic chip called μFILMS. In the talk, we show our initial results, chiefly construction of the μFILMS chip using additive manufacturing, analysis of the output signals and biochemical/mechanical adjustment of the chip parameters to optimize the visualization and/or characterization of the films. The construction of the μFILMS chip consists of a Transwell® static membrane placed between two glass slides, forming two distinct

438444 Photocatalytic Methanol Reforming on TiO2

Abstract:

Hydrogen gas has great potential for use in transportation fuels and industrial applications, but as it is currently produced commercially from steam reforming of unrenewable natural gas. An alternative source of hydrogen could be accessed from the reforming of sustainably-grown biomass. Through the use of a photocatalyst, biomass can be decomposed, producing clean-burning fuel. Titanium dioxide (TiO_2) is a metal oxide semiconductor of particular interest in photocatalysis due to its activity under ultraviolet illumination and its stability under catalytic reaction conditions. Ordinary TiO_2 (anatase, rutile) only absorbs UV wavelengths. Since a much larger fraction of solar irradiation is in the visible range, we aim to modify TiO_2-based photocatalyst to interact with visible light. Supporting metal nanoparticles on the surface of TiO_2 is known to cause a red shift in the absorbance of TiO_2. We have supported copper, nickel, silver, platinum, and palladium on TiO_2 through insipient wetness impregnation. The metal-modified TiO_2 was tested in two types of in situ reactors; one inside a simple kinetic reactor, and the other in a UV-Visible Spectrophotometer. The samples were then reacted under UV and visible light. Gaseous mixtures of methanol, oxygen and steam were passed over the catalysts while the temperature of the reaction was externally controlled. The products of the reaction were identified by gas chromatography. The addition of each of the metals shifted the absorbance of the TiO_2 to longer wavelengths. Future work includes further characterization of the materials, as well as varying the particle size of the metals and the support, and doping the support with hydrogen, carbon, and nitrogen.

438531 Room Temperature Shape Memory Polymers

Abstract:

Temperature controlled shape memory polymers can be synthesized via a facile method using long chain fatty acid and cross-linked natural rubber. For this experiment, rubber was swollen in liquid stearic acid (75 °C). Different weight percent loadings of acid were achieved by differing swelling times in one hour increments from 1 hour to 5 hours. The normal shape memory cycle consist of increasing the temperature of the swollen polymer above the melting point of the acid, stressing the polymer, then immediately quenching the heat in room temperature water. The rubber will hold stress without relaxing until heated above the melting point of stearic acid again, thus returning the rubber to its original state. Obtaining shape memory behavior at temperatures lower than 75 °C is desirable to mitigate safety concerns and reduce heating required, improving process economics. Investigation of the room temperature shape memory behavior was conducted through modifying the shape memory cycle. It was found that stearic acid swelled natural rubber exhibits a 30% fixity over a 24 hour period. The stress can be relaxed by heating the swelled rubber above the melting point of stearic acid.

Conceptual Process Design and Economic Analysis for Biomass Conversion Technologies to Valuable Chemicals

Yonggang Gao, Yongliang Chua, Pavan Kumar Naraharisetti, Daniel Gait Anak Kumbang, Tze Yuen Yeo, Arief Adhitya, Paul Sharratt, Jie Bu**

*Institute of Chemical and Engineering Sciences, A*STAR (Agency for Science, Technology and Research), Singapore*

Abstract

To quickly and reliably evaluate the economic potential of a proposed process in the early stage of process development, we have developed the iPEAD (Integrated Preliminary Economic Analysis and Design) methodology, which allows economic analysis of a process based on laboratory experiment results. In this paper, after the successful validation of iPEAD, we employ iPEAD for economic analysis of two biomass conversion processes: the chemical process of empty fruit bunch (EFB) to levulinic acid and the bioprocess of EFB to lactic acid. The minimum selling prices of levulinic acid and lactic acid are found to be 43% and 16% lower than market prices in 2013, respectively. Thus, it is concluded that the two processes are suitable to scale up to pilot plants for further investigation on process commercialization.

Introduction

Economic evaluation is a critical step for building, expanding or retrofitting a bio- or chemical process plant, since economic feasibility is a key factor for investors to invest in a project. In the lifecycle of product development and commercialization, the chances of commercialization are only 1 to 3% at the research stage, 10 to 25% at the development stage, and 40 to 60% at the pilot plant stage (1, 2). Thus, it is crucial to be able to make rapid economic assessment of major design decisions in the early stages of process development to avoid waste of funds and manpower. Furthermore, such assessment may identify both economic and technological challenges that are present in a given process and help scientists focus on areas that have the greatest impact. In light of this necessity to quickly visualize the economic position of a proposed process, the Process Engineering, Analysis and Design (PEAD) group at the Institute of Chemical and Engineering Sciences (ICES) has developed the iPEAD (Integrated Preliminary Economic Analysis and Design) methodology, which allows economic analysis of a process based on laboratory experiment results. In this paper, we employ iPEAD for economic analysis of two bio- and chemical processes of biomass conversion to valuable chemicals.

Using renewable natural resources to produce valuable chemicals has been attracting much attention as consumption of non-renewable crude oil becomes a growing environmental concern. In this study, two ICES-developed bio- and chemical processes are analyzed using iPEAD. The first process is levulinic acid production from empty fruit bunch (EFB) of palm oil by chemical conversion technology. The second process is lactic acid production from EFB by biological conversion technology (3). The objectives of the economic studies are (a) assess economic viability of the process based on preliminary experimental data; (b) identify the expensive process steps or factors resulting in high cost; (c) provide reasonable improvement suggestions or alternative routes/steps to reduce the cost.

iPEAD Methodology

In the iPEAD methodology, economic analysis of a proposed process is done by calculating the minimum selling price of the product(s). If the minimum selling price is lower than the market price, the process is considered to be economically viable. The basis for iPEAD analysis is laboratory experiment data. The iPEAD methodology consists of the following steps.

- *Collection of laboratory experiment data.* The information required usually includes types and quantities of starting materials, reaction conditions, residence time, conversions and yields, output composition and quantities, etc.
- *Flow sheeting.* Based on the laboratory experiment data, mass/energy balance, and fundamental engineering principles, a flow sheet is drawn by identifying the major equipment involved in the process, i.e. reactors, separation equipment (columns, filters, etc.) and heat transfer equipment.
- *Calculation of CapEx and OpEx.* From the flow sheet, the major equipment involved is sized according to a set capacity. In iPEAD, shortcut methods derived from authoritative recourses for sizing and costing various unit operations are employed to get a quick estimate of the major equipment cost (MEC)(4). Capital expenditure (CapEx) is derived from MEC by the use of Lang factors (5). In iPEAD, we finalize CapEx = 5 × MEC. The Chemical Engineering Plant Cost Index (CEPCI) is used to reflect cost changes over time (6). Operational expenditure (OpEx) consists of the cost of raw materials, consumables and energy in the form of electricity and heat.
- *Calculation of minimum selling price.* Cost structure for the minimum selling price is shown in **Figure 1**. The various cost items can be related to either CapEx or OpEx through factors (2). Overall, the minimum selling price is estimated as OpEx + CapEx/3.

For easy and practical use, iPEAD has been implemented as a software tool in Microsoft Excel with VBA macros.

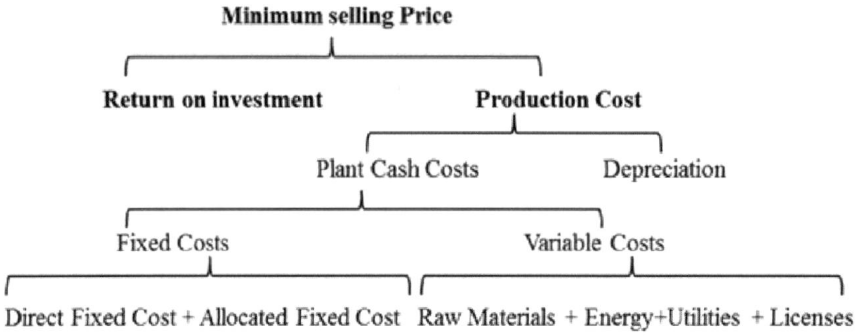

Figure 1. Cost structure of minimum selling price

iPEAD Validation

For validation, we have used iPEAD to estimate the CapEx of six processes from technical reports by professional and commercial agencies (**Table 1**). Results show a difference of -9.5% to 24.8% between the CapEx estimated by iPEAD and the technical reports, which is within the expected accuracy range (±50%) in the early stage of process development (7). This validates that iPEAD is sufficiently qualified and reliable for preliminary economic analysis.

Table 1. CapEx comparison between iPEAD and technical reports (estimation base year: 2013)

No.	Process	CapEx Reported, $ (million)	CapEx by iPEAD, $ (million)	Deviation, %
1	Steam methane reforming (SMR) for hydrogen (8)	528.75	659.83	+24.8
2	Olefins Conversion Technology (OCT) for propylene (9)	164.16	179.41	+9.3
3[a]	Levulinic acid production from paper sludge (10)	52.25	47.30	-9.5
4[b]	Gasoline production from wood (11)	142.72	138.09	-3.3
5[c]	Biofuel production – syngas cleaning (12)	15.77	16.08	+2.0
6	Biofuel production – fuel synthesis (12)	105.94	97.49	-8.0

a Not include the cost of waste boiler and hydrogenation of levulinic acid
b Sum of main equipment without heat exchangers
c Not include the cost of high pressure amine system

Economic Studies of Biomass-to-Chemical Processes

Southeast Asia produces over 80% of the world's palm oil, thus extraction of palm oil produces a mass of EFB, lignocellulose-rich biomass. ICES researchers have successfully developed chemical and biological processes to convert EFB to valuable chemicals: levulinic acid and lactic acid. We used iPEAD to analyze these two processes. For economic analysis, from lab-scale data the processes are scaled up to 100,000 ton of production per year.

EFB to Levulinic Acid

Levulinic acid is produced by EFB conversion via thermo-chemical technology. The process flow sheet is shown in **Figure 2** and the iPEAD results for CapEx and OpEx are shown in **Table 2**. It is found that Distillation 1 and 2 costs account for 67% of total CapEx. This is due to the large amount of extractive solvent used in liquid-liquid extraction, which needs to be recovered in downstream process. The large amount of solvent recovery also leads to high heating cost, accounting for 64% of total OpEx. Based on these economic analysis results, we suggested researchers to cut down the cost by reducing the amount of extractive solvent used in upstream process. Simultaneously, we found that raw material (EFB) accounts for 17% of total operational costs, thus we suggested to increase the product yield, which would reduce not only the amount of raw materials, but also CapEx per ton of levulinic acid. At the current stage, the minimum selling price of levulinic acid is $1435, much lower than the market price, $2500 (13).

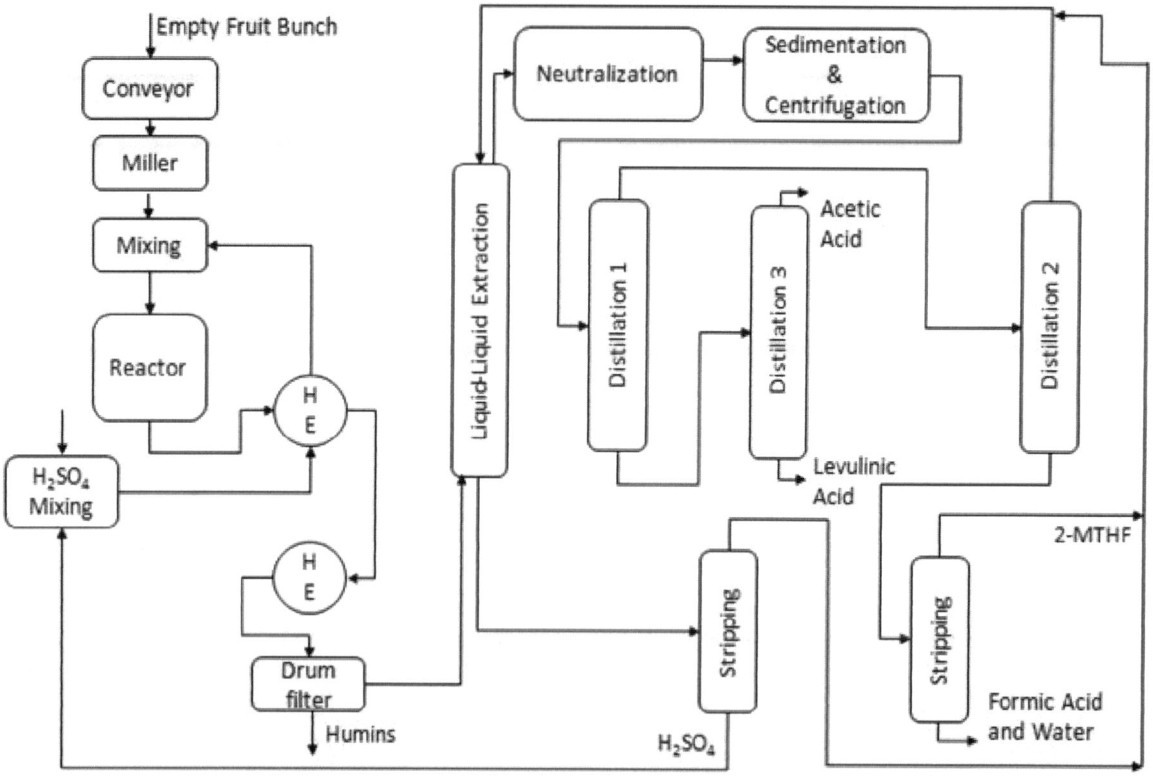

Figure 2. Flow sheet of EFB to levulinic acid process (HE: Heat Exchanger; 2-MTHF: methyltetrahydrofuran; H_2SO_4: Sulphuric Acid)

Table 2. CapEx and OpEx of EFB to levulinic acid process (estimation year 2013; 100,000 ton of levulinic acid/year)

	M$	**$/ton product**	**% of CapEx**
CapEx	327.49	3274.90	100.00%
Reactor	4.81	48.07	1.47%
Heat exchangers	8.09	80.88	2.47%
Drum filter	45.21	452.08	13.80%
LLE tank	32.58	325.76	9.95%
Distill. 1	90.17	901.69	27.53%
Distill. 2	128.81	1288.13	39.33%
Distill. 3	6.96	69.56	2.12%
Others	10.88	108.75	3.32%

	Unit Price $	**$/ton product**	**% of OpEx**
OpEx		343.55	100.00%
Energy Costs		*258.62*	
Electricity per kWh	0.1	33.38	9.72%
Heating per GJ	4.00	220.17	64.09%
Misc		5.07	1.48%
Raw Material Costs		*84.93*	
EFB per ton	7	59.65	17.36%
Water makeup per ton	0.58	3.89	1.13%
Acid makeup per ton	50	19.89	5.79%
Solvent makeup per kg	1.5	1.50	0.44%

EFB to Lactic Acid

Lactic acid is produced by EFB conversion via biological technology. B. coagulans JI12 ICES patented bacterial strain was used in the biological process (3). The process flow sheet is shown in **Figure 3**. The CapEx and OpEx estimated by iPEAD are shown in **Table 3**.

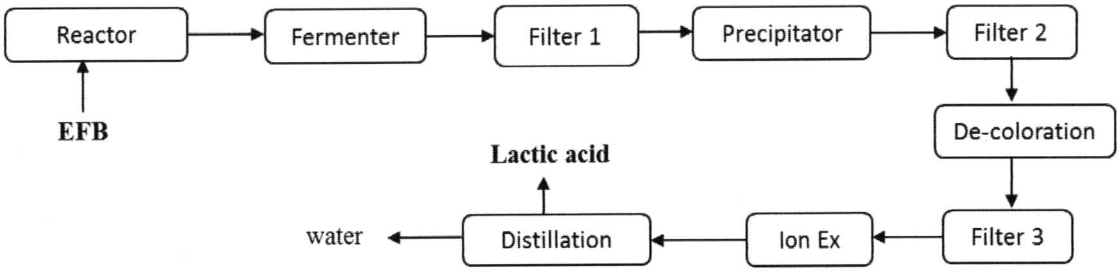

Figure 3. Flow sheet of EFB to lactic acid process

Table 3. CapEx and OpEx of EFB to lactic acid process (estimation year 2013; 100,000 ton of lactic acid/year)

	M$	**$/ton product**	**% of CapEx**
CapEx	265.47	2654.70	100.00%
ION-Exchangers	26.46	264.6	9.97%
Acid Hydrolysis	5.16	51.6	1.94%
Fermenters	165.15	1651.5	62.21%
CaSO4 ppt reactor	5.97	59.7	2.25%
Decolor reactor	32.85	328.5	12.37%
Filtration	11.45	114.5	4.31%
Distillation	18.43	184.3	6.94%

	Unit Price $	**$/ton product**	**% of OpEx**
OpEx		788.79	100.00%
Energy Costs		*80.93*	
Electricity per kWh	0.1	25.00	3.17%
Heating per GJ	4.00	55.93	7.09%
Raw Material Costs		*707.86*	
EFB (undried fibers)	40	106.88	13.55%
Sulfuric acid	100	45.97	5.83%
Phosphoric acid	600	35.78	4.54%
Hydrated lime	100	57.49	7.29%
Carbon	100	60.37	7.65%
Enzyme	2000	231.94	29.40%
Sugar 1	500	33.13	4.20%
Sugar 2	500	33.13	4.20%
Yeast extract	2000	26.51	3.36%
Calcium carbonate	100	3.98	0.50%
Salts	400	3.71	0.47%
ION-Exchanger resin	3000	66.47	8.43%
Resin regeneration:HCl	100	0.50	0.06%
Resin regeneration: NaOH	400	2.00	0.25%

It is found that the fermenter cost accounts for 62% of total CapEx. This is due to the large amount of water used in fermentation. Thus, we suggested using less water/fermentation time to reduce the volume of fermenters. Simultaneously, we found that raw materials EFB and enzyme account for 14% and 29% of total OpEx costs, respectively, thus we suggested increasing the product yield and using cheaper enzyme or reducing enzyme amount. The yield increment can reduce the CapEx and raw materials costs per ton of lactic acid. At the current stage, the minimum selling price of lactic acid is $1674, lower than the market price, $2000 (14).

Conclusions

We have developed iPEAD as a methodology for preliminary economic analysis based on laboratory experiment results. Validation was done through comparison with published technical reports and iPEAD is proven to be sufficiently competent and reliable for economic analysis at the early stage of development. iPEAD has successfully been applied to analyze two EFB conversion processes. Analysis results help to identify opportunities to further reduce costs through improving technical issues and expensive process aspects. iPEAD methodology makes the preliminary economic analysis less dependence or even no dependence of Aspen Plus software and professional modeling skills.

References

1. B. T. Frohlich (1999), "Considerations for Developing In-House Capabilities for Cost of Goods Estimation," *Biopharmaceutical Process Economics and Optimization*, September 30-Otober 1.
2. J. M. Douglas (1988), "Engineering Economics," *Conceptual Design of Chemical Processes*, International ed., New York: McGraw Hill, pp. 23-71.
3. L.D. Ye, M.S.B. Hudari, Z. Li, J.C. Wu (2014), "Simultaneous Detoxification, Saccharification and Co-fermentation of Oil Palm Empty Fruit Bunch Hydrolysate for L-lactic Acid Production by Bacillus Coagulans JI12," *Biochemical Engineering Journal*, 83, pp. 16-21.
4. M. Peters, K. Timmerhaus and R. West (2002), *Plant Design and Economics for Chemical Engineers*, 5th ed., McGraw-Hill.
5. R. K. Sinnott (1999), "Costing and Project Evaluation," *Coulson and Richardson's Chemical Engineering Design*, 3rd ed., Oxford: Butterworth Heinemann, pp. 242-282.
6. G. D. Ulrich and P. T. Vasudevan (2009), "Capital Costs Quickly Calculated," *Chemical Engineering*, 116, pp. 46-52.
7. J. C. Jr. Lagace (2006), "Making Sense of Your Project Cost Estimate," *Chemical Engineering*, 113, pp. 54-58.
8. Inc. Nexant (2006), "Equipment Design and Cost Estimation for Small Modular Biomass Systems," *Synthesis Gas Cleanup and Oxygen Separation Equipment*, U.S. Department of Energy, Oak Ridge, TN. From www.nrel.gov/docs/fy06osti/39943.pdf
9. Sol. Intratec (2013), *Technology Economics Propylene via Metathesis, Intratec Solutions*, Houston, TX. From http://www.intratec.us/publications/propylene-production-via-metathesis.
10. Inc. Biometrics (2002), *Commercialization of the Biofine Technology for Levulinic Acid Production from Paper Sludge*, Biometrics Inc., Waltham, MA. From http://www.osti.gov/scitech/biblio/771246
11. S. D. Phillips, J. K. Tarud, M. J. Biddy and A. Dutta (2011), *Gasoline from Wood via Integrated Gasification, Synthesis and Methanol-to-Gasoline Technologies*, National Renewable Energy Laboratory, Golden, CO. From www.nrel.gov/docs/fy11osti/47594.pdf

12. R. M. Swanson, J. A. Satrio, R. C. Brown, A. Platon and D. D. Hsu (2010), *Techno-Economic Analysis of Biofuels Production Based on Gasification*, National Renewable Energy Laboratory, Golden, CL. From www.nrel.gov/docs/fy11osti/46587.pdf
13. Carbolea (2014), "Studies on the Economics of Biorefining Technologies," *Carbolea*. From http://www.carbolea.ul.ie/economics_lit.php.
14. ICIS Pricing (2013), "Lactic Acid," *Reed Business Information*, Surrey.

421425 Scenarios Simulation and Forecast of Urban Economy-Resources-Environment System By Using System Dynamics Model

Abstract:

The coordinated development of Economy-Resources-Environment (ERE) is a key component of sustainable development. Based on the connotation and framework of sustainable development of urban ERE, this paper employs the theory and methodology of System Dynamics (SD) to elucidate the economic growth constrained from urban resource depletion and environmental degradation. Herein, resource, environment, and technology subsystems are combined with economy subsystem into a Cobb-Douglas production. Meanwhile, five scenarios, named current, resources, environment, technology and economy, are designed to compare the simulation results, so that the effects of different scenarios on the urban ERE can be emphasized. The results demonstrate that the current development of Chongqing is unsustainable, and the optimization of any one subsystem cannot realize the sustainable development. It turns out that the transformation of the economic development mode is the only way to sustainable development in Chongqing, and feasible transformation ways of the economic development mode have been put forward.

427677 Energy Integration Alternatives for Effective Use of Waste Heat Associated with Flaring during Abnormal Situation

Abstract:

TG | TOPICAL CONFERENCE: INNOVATIONS OF GREEN PROCESS ENGINEERING FOR SUSTAINABLE ENERGY AND ENVIRONMENT

Energy Integration Alternatives for Effective Use of Waste Heat Associated with Flaring during Abnormal Situation

Monzure-Khoda Kazi, Fadwa Eljack*

Department of Chemical Engineering, College of Engineering,

Qatar University, P.O. Box-2713, Qatar

Email: Fadwa.Eljack@qu.edu.qa

This work reports a multi-objective optimization frame work for new/modified design solutions for energy integration to manage flares from uncertain sources during abnormal situation. It also suggests some energy integration alternatives for waste streams recovery, for combining heat and power through process integration and for flare gas utilization. This kind of robust optimization framework and energy integration alternatives are very important for utilizing unused resources of the process and for flare gas reduction. Although, governments and companies have had success in reducing flare gas with significant investments; global gas flaring has remained largely stable the past fifteen years in the range of 140-170 billion cubic meters (BCM) which is 5% of global gas production (Kazi et al., 2015).

Therefore, much attention is still needed to mitigate and manage industrial flares by innovating green process engineering for sustainable energy and environment.

Here, two sustainable energy integration alternatives are proposed and harnessed properly with process for flare management during abnormal situation. First one is cogeneration (COGEN) unit which has become a mainstream practice in different industries due to its high economic and energy-savings potentials. It can produce simultaneous heat and power from a single fuel source. If COGEN unit is designed properly some part of the flare streams can be used as fuel in the boiler instead of fresh fuel like NG. This will reduce the fresh fuel cost and will also increase the energy efficiency of the process itself. Moreover, it will reduce the carbon footprint from compensated fresh fuel. Another energy integration tool is thermal membrane distillation (TMD) unit, which is an emerging technology in the area of high-purity separation especially in water treatment (Elsayed et al., 2013). In TMD, low level heat like waste heat from process or COGEN unit can be used to create the driving force to separate pure water from the raw/waste water. Moreover, it needs small floor area for installation and it is modular in nature. By integrating TMD with the process, the utility costs and the waste water treatment cost can be reduced significantly. Produced water from TMD can be recycled in the process or sold to external users. Using cogeneration (COGEN) and thermal membrane distillation (TMD) unit with industrial processes can offer a potential scope of process and energy integration and can also offer the maximum utilization of flared gas streams.

In this work, an ethylene base case study is solved to illustrate the applicability of the proposed optimization framework and the implementation of the proposed energy integration tools for flare management. In final output, the optimization framework provides several Pareto fronts to the policy makers. Therefore, the end user is able to make an informed choice of suitable design, operation conditions and process performances depending on economic, energetic and/or environmental considerations. The results of different scenarios demonstrate the benefits of integrating TMD and COGEN unit with the process plant for flare mitigation approach. It will ensure the maximum use of flared gas streams, reduction of cooling utility load and the creation of recycle/reuse opportunities for the treated waste water.

Reference

Elsayed, N. A., Barrufet, M. A., & El-Halwagi, M. M. (2013). Integration of Thermal Membrane Distillation Networks with Processing Facilities. Industrial & Engineering Chemistry Research, 53, 5284-5298.

Kazi, M.-K., Mohammed, F., AlNouss, A. M. N., & Eljack, F. (2015). Multi-objective optimization methodology to size cogeneration systems for managing flares from uncertain sources during abnormal process operations. Computers & Chemical Engineering, 76, 76-86.

428647 A Case for Integrated Process Engineering

Abstract:

As competition gets severe, innovation in the whole value chain of a product life cycle is the lifeblood of a technology. Unfortunately, the execution at almost every step of the life cycle in creating and commercializing an innovation currently is mostly sequential. The lack of interactions among phases and sections in the value chain leads to losses of budget, opportunity and competitive advantages, in particular in breakthrough technology development for sustainable energy and environment, which is a key factor hindering the success of technology transfer from science to products. This paper discusses some typical applications of integrated process engineering and economic analysis.

431829 The Role of Carbon Dioxide As a Solvent for Greener Biodiesel Production over a Heterogeneous Catalyst

Abstract:

Within the field of alternative energies, fuels produced from biological feedstocks have great potential to replace petroleum-based products. In order to sustainably produce fuels and materials, biorefineries must be designed for and operated with sustainable processes. Compressed carbon dioxide – subcritical, supercritical, and CO_2 expanded liquids – can potentially be used in a biorefinery as a green solvent. Key processes that may be improved via the use of CO_2 are extraction (including fractionation and refinement) and reaction. In this presentation, supercritical CO_2 ($scCO_2$) is shown to be an effective solvent for the selective conversion of lipids into biodiesel in combination with a heterogeneous catalyst. Successful transesterification of triglycerides into fatty acid methyl esters is demonstrated via a CO_2-expanded liquid phase. The role of CO_2 in reaction moderation is explored via kinetics with potential implications for selectivity and system optimization. While pitfalls do exist for the application of compressed CO_2 in commercial production, this versatile compound holds great promise for utilization as a green solvent in a biorefinery setting.

436989 New Process for Recovery of Renewable Energy and Chemicals from Black Liquor – A Review

Abstract:

In the pulp and paper industry, large quantities of forest biomass are being used. The by-products or residues which result include black liquor, bark, and forest logging residues. These can be used for energy purpose to produce electricity, heat, steam, and biofuels. The kraft pulping process accounts for almost 60% of all pulp production. Black liquor from kraft process represents a potential energy source of 250-500 MW per mill. This paper describes an exciting emerging technology that uses low temperature steam re-forming instead of traditional combustion to recover energy and chemicals from black liquor boosts high thermal efficiency, low emissions (including particulate matters and odorous TRS gases), and low power consumption. This versatile process, a very good example of pollution prevention at the source through process changes, is now available for commercial scale processing of black liquors generate at the pulp and paper industry. The black liquor steam reforming process utilizes indirect heating of a steam fluidized bed of sodium carbonate solids to recover cooking chemical and energy. Black liquor is sprayed directly into the bed where the liquor droplets uniformly coat the bed solids, resulting in high rates of heating, pyrolysis and steam reforming. The process steam reforms black liquor in the absence of oxygen, producing a hydrogen-rich, medium-Btu gas with a heating value of 300-350 Btu/SCF. Bed temperatures are maintained at 1060 to 1150 ^{0}F (565-620 ^{0}C), thereby avoiding liquid smelt formation and the associated smelt-water explosion hazards. In this dry (no smelt) recovery process, the sodium sulfate in the liquor is reduced by reaction with the steam-reforming products, principally carbon monoxide and hydrogen. The reduced sulfur form is unstable in the reformer environment and decomposes rapidly to gaseous hydrogen sulfite (H_2S) and solid sodium carbonate (Na_2CO_3). The product gas is scrubbed with a solution derived from the sodium carbonate bed solids to produce green liquor for re-use in the pulping cycle.

Heat required to reach operating temperature and for the endothermic steam-reforming reactions is supplied by heat exchangers immersed in the fluidized bed. The heat exchangers consist of bundles of pulsed heater resonance tubes. A portion of the product gas is burned in the pulsed heaters to supply the necessary heat, thus making the steam reformer self-sufficient on its own fuel. Pulsations in the resonance tubes produce a gas-side heat transfer coefficient which is up to five times greater than a conventional fire-tube heater. This efficient heat transfer reduces the size and cost of the heat exchangers and reformer vessel. The hot combustion gases leaving the pulsed heaters are sent to a waste heat boiler to generate steam and to preheat the pulsed heater combustion air. The product gas from the reformer is passed through a cyclone to remove particulate matter, then cooled in a waste heat steam generator, scrubbed and quenched.

By separating the sodium and sulfur elements, this process provides an imaginative and chemically attractive alternative to conventional kraft recovery. The process separates the individual unit operations of the conventional recovery boiler and allows independent control of each operation under its own optimized conditions, producing a well-controlled, more efficient modular process plant. It is applicable to both incremental and full-scale production facilities. The process also provides attractive energy alternatives, since the product gas is compatible with combined-cycle cogeneration operations.

434099 Study of Iron Oxide Based Mixed Oxygen Carriers for in-Situ Tar Cracking in Biomass Chemical Looping

Abstract:

A chemical looping process that uses biomass as the feedstock could prove to be a highly economical, environmental and sustainable means for electricity and/or chemical production with net negative carbon emissions. A chemical looping system converts carbonaceous fuels, such as biomass, using metal oxide based oxygen carriers circulating between three reactors - a reducer, an oxidizer and a combustor – to produce heat and high purity hydrogen with simultaneous carbon dioxide (CO_2) sequestration. In the reducer, the biomass fuel is converted to CO_2 and H_2O by oxygen transfer from the oxygen carriers. The reduced oxygen carriers are then partially oxidized by steam in the oxidizer producing high purity hydrogen and subsequently fully oxidized by air in the combustor. The energy generated due to the exothermic reactions in the combustor is sufficient to compensate for the heat required to perform the endothermic reactions in the reducer and to produce electricity via steam generation.

Tar evolved during biomass conversion in the reducer mainly consists of oxygen containing heavy hydrocarbons and complex polycyclic aromatics. It is a major concern as it can hinder the continuous operation and performance of a biomass chemical looping (BCL) system. Hence, it is important to design oxygen carriers that have the ability to eliminate the tars in the reducer. While Fe_2O_3 is an ideal candidate for oxygen carrier material for chemical looping processes, in this study, different tar cracking catalysts (Olivine, Dolomite, K_2CO_3 and NiO) have been used in conjunction with Fe_2O_3 to enhance the carriers' ability to crack tar. Thermogravimetric Analysis (TGA) has been used to study the ability of the oxygen carriers to undergo multiple redox cycles. A series of fixed bed experiments have been used to crack biomass-derived tar with the mixed oxygen carriers. The product gaseous species have been identified using Gas Chromatography–Mass Spectrometry (GC-MS) and fresh and reacted solids have been characterized using X-Ray Diffraction (XRD). Addition of NiO to the Fe_2O_3 based oxygen carrier was observed to greatly improved its tar cracking ability and maintained its redox reactivity over multiple cycles. The results from this study will help in the development of oxygen carriers towards designing an optimal BCL system.

427895 Synthesis of SBA-16 Supported Cesium Salt of Iron Substituted Lacunary Keggin Type Polyoxometalate Based Catalyst for Selective Thermal Catalytic Cracking of Low Density Polyethylene

Abstract:

Polyethylene is a non-biodegradable polymer which may cause serious environmental issues, but its cracking can produce fuel oil and may solve the exhausting fuel issue. Thermal degradation of waste plastics into fuel oil is one of the most promising methods for conversion of the plastic wastes into fuel. However, in catalytic cracking of polyethylene wastes more solid residue with wax is achieved. And the liquid product contains a range of hydrocarbons which affect the nature as well as yield of the fuel fractions. The development of a suitable catalyst with high selectivity is highly desirable. Polyoxometalates (POMs) are class of molecularly defined inorganic metal-oxide clusters formed from early transition metals (vanadium, tungsten etc.). Their use as catalysts can result in selective recovery of useful hydrocarbons from waste polyethylene. This research is based on synthesis of SBA-16 supported cesium salt of iron substituted lacunary keggin type polyoxometalate. The synthesized material was characterized using FTIR, TGA-DSC, XRD and SEM-EDX to investigate their structure and evaluate their properties. The material was thermally stable above 500 ^{o}C and therefore used as catalyst for thermal cracking of low density polyethylene. The cracking temperature was evaluated by TGA-DSC of physical mixture of POM and polyethylene and experimental conditions were designed accordingly. The catalytic cracking was carried out in a batch reactor and the product was analysed by GC-MS. The results showed that the synthesized catalysts exhibited excellent efficiency in conversion of maximum polyethylene into liquid oil. This research will contribute a lot to solve the natural resource depletion issues and environmental issues.

429411 Energy and Carbon Foot Print Analysis of the Production of a Liquid Organic Hydrogen Carrier

Abstract:

In a future energy system based on renewable energies, energy carriers and storage systems are required to store excess energy for times with low provision of energy from renewable resources. In this context, hydrogen is considered as a suitable energy carrier although the efficient, dense and secure storage of hydrogen is still a challenge.

A promising approach to overcome this limitation is the use of liquid organic hydrogen carriers (LOHCs). LOHCs are unsaturated organic compounds which can store hydrogen by reversible hydrogenation (e.g. dibenzyltoluene). The hydrogenated form of the LOHC exhibits high thermodynamic stability, enabling storage at ambient conditions. In contrast to the application of solid chemical carrier substances e.g. metal hydrides, the handling and transportation of the LOHC rather easy due to the liquid nature of the carrier substance.

A lot of research was performed concerning the LOHC storage concept in terms of applicability and efficiency but only little is known about the energy demand and the ecological impact of the production of the carrier as such. Especially the carbon foot print of an energy carrier substance is important to ensure the overall ecological benefit of the storage concept.

In this contribution, we present a process concept for the production of the LOHC dibenzyltoluene on technical scale. Based on the process concept, the heat and electricity demand is determined and potential energy savings are identified by a pinch analysis. Moreover, a carbon foot print analysis is performed to reveal the ecological impact related to the production of the LOHC.

First results show that the main energy demand of the overall process is linked to the product purification steps. Especially for low conversions in the synthesis reactor this energy demand increases remarkably. By rigorous heat coupling of the overall process, the overall heat demand can be lowered by about 35 %.

428468 Process Synthesis of Flexible and Sustainable Processes Consisting of Membrane Separations and Reactive Distillation for the Production of Biodiesel

Abstract:

Due to global competition and an increasing demand for reduced emissions the development of sustainable and efficient processes based on renewable resources becomes more and more important. Furthermore, the fluctuating qualities of renewable feedstock necessitate highly flexible processes. Fatty esters are promising new building blocks for the chemical industry and can be produced by the transesterification of virgin vegetable oils. However, nowadays fatty esters are mainly used as biodiesel, an alternative fuel in future based on renewable resources [1,2]. Due to the high raw material costs for the virgin vegetal oils [3], no economic production is possible. A promising approach to lower the production costs is the application of cheap low-quality oils, such as waste cooking oil or crude non-edible oil [4,5]. These low quality oils contain a high amount of free fatty acids, which lead to undesired side reactions in the alkaline catalyzed transesterification reaction [6,7]. Consequently, new process concepts are necessary to enable the application of low-quality oils as a feedstock.

In order to achieve an efficient processing of the low-quality oils this study investigates intensified process concepts, such as reactive distillation and membrane-assisted reactive distillation. One possibility to reduce the free fatty acid content in the low-quality oils is the acid catalyzed esterification of the free fatty acids with an alcohol. Since this reaction is equilibrium limited the application of reactive distillation is favorable to enable higher conversion. Furthermore, it is indicated that the transesterification of the oils could take place simultaneously. To further enhance the conversion in the reactive distillation the combination with vapor permeation is promising. The selective removal of the by-product water from the condensate stream facilitates a shift of the reaction equilibrium towards the product side. Another possibility to reduce the free fatty acid content is the separation of the free fatty acids from the low-quality oils by means of organic solvent nanofiltration. The application of organic solvent nanofiltration enables an energy efficient separation due to the mild operation conditions, which also ensure consistent quality of the fatty acids and oils. The coupling of organic solvent nanofiltration with reactive distillation provides the potential to an even more energy efficient process. Besides the processing of the low-quality oils, the profitability of the production process might be increased because of the generation of new products, as e.g. the free fatty acids, which can be used as building blocks in the chemical industry.

The current study investigates the process synthesis of different flow sheet configurations of reactive distillation, vapor permeation and organic solvent nanofiltration. Therefore, models for each unit operation are developed and validated with own experimental results and literature data. Hence, the potential of the different intensified process configurations is evaluated and it is determined if they are capable to handle fluctuating feedstock compositions. Finally, the most promising configurations are identified with regard to product specifications, energy demand and production costs.

References:

[1] Demirbas, A., "Biodiesel: A realistic fuel alternative for diesel engines," Springer, London (2008).

[2] Santacesaria, E., Vicente, G.M., Di Serio, M., and Tesser, R., "Main technologies in biodiesel production: State of the art and future challenges," *Catalysis Today*, 195, pp. 2–13 (2012).

[3] Lim, S., and Teong, L.K., "Recent trends, opportunities and challenges of biodiesel in Malaysia: An overview," *Renewable and Sustainable Energy Reviews*, 14, pp. 938–954 (2010).

[4] Phan, A.N., and Phan, T.M., "Biodiesel production from waste cooking oils," *Fuel*, 87, pp. 3490–3496 (2008).

[5] Yaakob, Z., Mohammad, M., Alherbawi, M., Alam, Z., and Sopian, K., "Overview of the production of biodiesel from Waste cooking oil," *Renewable and Sustainable Energy Reviews*, 18, pp. 184–193 (2013).

[6] Freedman, B., Pryde, E.H., and Mounts, T.L., "Variables affecting the yields of fatty esters from transesterified vegetable oils," *Journal of the American Oil Chemists Society*, 61, pp. 1638–1643 (1984).

[7] Kumar Tiwari, A., Kumar, A., and Raheman, H., "Biodiesel production from jatropha oil (Jatropha curcas) with high free fatty acids: An optimized process," *Biomass and Bioenergy*, 31, pp. 569–575 (2007).

418941 Water Solubility and Micellar Formation of Imidazolium and Ammonium Ionic Liquids

Abstract:

A number of proposed applications for ionic liquids involve ionic liquid/water interfaces; such as, chemical separations or drug delivery systems. Therefore an understanding of the solubility and micellar behavior of ionic liquids in an aqueous environment is critical fundamental knowledge. The long term goal of our study was the utilization of ionic liquids for the delivery of biological active agents. We chose the anion, bis(trifluoromethanesulfonyl)imide (Tf_2N) since it promotes water stability and forms water immiscible RTILs. The study then paired the Tf_2N-anion with three different classes of cations. By adding apolar substituents ($-CH_2-$) to the cation, the size of the molecule increases in each RTIL class, rendering the RTIL less soluble in the water phase. The three classes of RTILs examined were 1-C_n-3-methylimidazoliums, C_n-trimethylammoniums (CTA), and branched ammoniums (BAM), with the "C_n" representing an alkyl functional group (propyl, butyl, etc.). For industrial use, CTAs are attractive, since they can be synthesized from inexpensive ammonium surfactants. However, large CTAs (twelve carbon chain and higher) are solids at ambient conditions. In contrast, BAMs have larger compounds that remain in the liquid state at ambient conditions. Nevertheless, each class has potential as a delivery agent by meeting the aforementioned criterion – namely, water immiscible and water stable at ambient temperatures. We used Total Organic Carbon (TOC) analysis to determine the RTIL content in water that is in equilibrium with the tested RTILs. Surface tension measurements of the RTIL containing water determined if RTIL micelles existed in the RTIL saturated water. We used Linear Free Energy Relationship (LFER) semi-empirical models to correlate the RTIL water solubility to the molecular size and structure of the RTIL cation. Combining the LFER results with surface tension measurements and thermodynamic calculations of sub-cooled RTIL fugacities allowed us to determine that micelle formation is not significant for the tested RTILs with molar volumes less than 350 cm^3/mol. We tested a total of 15 different RTILs with the water solubilities ranging from 10-2 to 10-4 moles-RTIL/L-water.

421632 Insights into Cellulose Dissolving in Ionic Liquids By Molecular Simulation

Abstract:

Insights into Cellulose Dissolving in Ionic Liquids by Molecular Simulation

Xiaomin Liu,[a] Yao Li,[a] Yuling Zhao,[b] Jianji Wang,[b] and Suojiang Zhang[a,*]

a State Key Laboratory of Multiphase Complex Systems, Institute of Process Engineering, Chinese Academy of Sciences, Beijing, 100190, China.

b Key Laboratory of Green Chemical Media and Reactions, Ministry of Education, School of Chemistry and Environmental Science, Henan Normal University, Xinxiang, Henan 453007, China.

Abstract

Ionic liquids (ILs) have been used in an enormous number of chemical and biological processes. In recent years, great progress has been made in the dissolution of cellulose with ILs. However, the mechanism of cellulose dissolution has not been clearly understood. [1,2]

Molecular simulations were performed to investigate the effects of cation and anion structures on cellulose dissolution in ILs. It was found that the presence of electron-withdrawing group in alkyl chain of the cation enhanced the interaction between the cation and cellulose by comparing the interaction energy between $[Amim]^+$ and $[C_3mim]^+$ with cellulose.[3] Effects of anionic structure on the dissolution of cellulose in ILs revealed that hydrogen bonds (HBs) were formed between anions of the ILs and hydroxyl protons of cellulose. we also investigated the effects of electronegativity of HB acceptor atoms, steric effect of alkyl chain length and electron-withdrawing group of the anions on their HB acceptor ability. [4] Furthermore, molecular simulations were performed to depict the cosolvent effect of cellulose dissolution in imidazolium-based ILs. DMSO partially breaks down the ionic association of $[C_4mim][CH_3COO]$ by solvation of the cation and anion, but no preferential solvation was observed. The dissociated anions would readily interact with cellulose to improve the dissolution of cellulose. [5] Besides, the spontaneous dissolution process was investigated through all-atom molecular dynamics simulations of cellulose bunche in EmimAc and EmimCl within 0.5~2μs. Synergistic effect of cations and anions was revealed by analysing the whole dissolving process in EmimAc. Initially, cations bind to the side face of cellulose bunch and anions insert into the cellulose strands to form HBs with hydroxyl groups. Then cations start to intercalate into cellulose chains due to their strong electrostatic interaction with the anions.

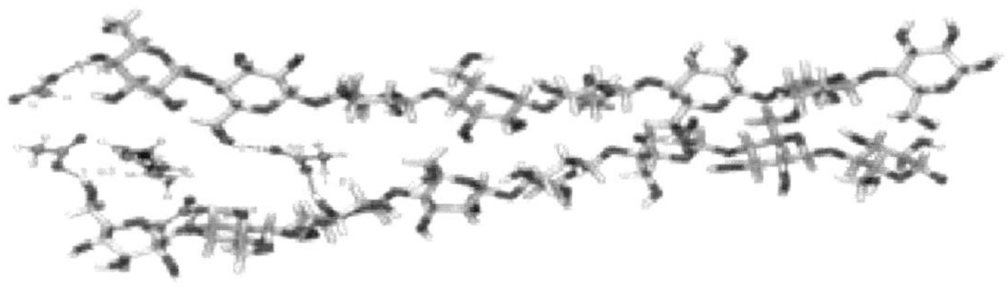

Figure 1. The spontaneous dissolution snapshot of cellulose in EmimAc.

References

1. S. Zhang, J. Sun, X. Zhang, et al, *Chem. Soc. rev.*, 2014, 43, 7838-7869.

2. K. Dong, Q. Wang, S. Zhang, *Sci. China Chem.*, 2015, 58,495-500.

3. Y. Zhao, X. Liu, J. Wang, S. Zhang. *ChemPhysChem*, 2012, 13, 3126-3133.

4. Y. Zhao, X. Liu, J. Wang, S. Zhang. *Carbohydr. Polym.*, 2013, 94: 723-730.

5. Y. Zhao, X. Liu, J. Wang, S. Zhang. *J. Phys. Chem. B*, 2013, 117, 9042-9049.

were formed between anions of the ILs and hydroxyl protons of cellulose. we also investigated the effects of electronegativity of HB acceptor atoms, steric effect of alkyl chain length and electron-withdrawing group of the anions on their HB acceptor ability. [4] Furthermore, molecular simulations were performed to depict the cosolvent effect of cellulose dissolution in imidazolium-based ILs. DMSO partially breaks down the ionic association of $[C_4mim][CH_3COO]$ by solvation of the cation and anion, but no preferential solvation was observed. The dissociated anions would readily interact with cellulose to improve the dissolution of cellulose. [5] Besides, the spontaneous dissolution process was investigated through all-atom molecular dynamics simulations of cellulose bunche in EmimAc and EmimCl within 0.5~2μs. Synergistic effect of cations and anions was revealed by analysing the whole dissolving process in EmimAc. Initially, cations bind to the side face of cellulose bunch and anions insert into the cellulose strands to form HBs with hydroxyl groups. Then cations start to intercalate into cellulose chains due to their strong electrostatic interaction with the anions.

Figure 1. The spontaneous dissolution snapshot of cellulose in EmimAc.

References

1. S. Zhang, J. Sun, X. Zhang, et al, *Chem. Soc. rev.*, 2014, 43, 7838-7869.

2. K. Dong, Q. Wang, S. Zhang, *Sci. China Chem.*, 2015, 58,495-500.

3. Y. Zhao, X. Liu, J. Wang, S. Zhang. *ChemPhysChem*, 2012, 13, 3126-3133.

4. Y. Zhao, X. Liu, J. Wang, S. Zhang. *Carbohydr. Polym.*, 2013, 94: 723-730.

5. Y. Zhao, X. Liu, J. Wang, S. Zhang. *J. Phys. Chem. B*, 2013, 117, 9042-9049.

421710 Novel CO2 Separation Process By Using Ionic Liquids

Abstract:

Novel CO_2 separation process by Using Ionic Liquids

Haifeng Dong, Suojiang Zhang*

Institute of Process Engineering, Chinese Academy of Sciences, 100190, Beijing, China

**Corresponding author: sjzhang@ipe.ac.cn*

As a new solvent or medium, the unique structures and properties of ionic liquids make them show significant advantages in the reaction and separation fields. That offers new opportunities for new technology development and the breakthrough of reaction conditions [1]. Previously, we have done some interesting work for understanding the fundamental nature of ILs and developing new gas separation process. New task-specific ILs for CO_2 separation were designed and synthesized by modifying the structure of the anion and cation of the ILs. The CO_2 capture capacity of dual amino-functionalized ionic liquids is up to 18.5 wt%, which is obviously higher than previously reported amino-functionalized ILs [2]. Ionic-fragment method and corresponding states principle were proposed to develop predictive models for thermodynamic properties and phase equilibria of ILs system [3]. The effect of small amount of water on CO_2 bubble behavior in ionic liquid was investigated. The results indicated that a small amount of water has significant influence on the bubble behavior and mass-transfer coefficient in ionic liquids [4].

Based on simulated flue gas, continuous experiments were completed to evaluate the ionic liquids performance in the laboratory by using gas separation unit (1.0Nm^3/h). Process simulation was also developed to optimize the CO2 removal performance using pure ILs and ILs mixture. The results showed that ILs process saved about 31% regeneration heat energy consumption and 13.5% total capture cost compared to the monoethanolamine process[5]. Next step, we hope to promote the engineering design and construction of demonstration equipments by cooperating with company.

Reference:

[1] X. P. Zhang, X. C. Zhang, H. F. Dong, Z. J. Zhao, S. J. Zhang, Y. Huang. Carbon capture with ionic liquids: overview and progress. ***Energy Environ. Sci.***, 2012, 5, 6668

[2] J. Z. Zhang, C. Jia, H. F. Dong, J. Q. Wang, X. P. Zhang, S. J. Zhang. A Novel Dual Amino- Functionalized Cation-Tethered Ionic Liquid for CO_2 Capture. ***Ind. Eng. Chem. Res.***, 2013, 52(17), 5835

[3] Y. Huang, H. F. Dong, X. P. Zhang, C. S. Li, S. J. Zhang. A new fragment contribution- corresponding states method for physicochemical properties prediction of ionic liquids. ***AIChE J.***, 2013, 59(4), 1348

[4] X. Zhang, H. F. Dong, D. Bao, Y. Huang, X. Zhang, S. J. Zhang. E ect of small amount of water on CO_2 bubble behavior in ionic liquid systems. ***Ind. Eng. Chem. Res.*** 2014, 53, 428-439.

[5] Y. Huang, X. P. Zhang, H. F. Dong, X. Zhang, S. J. Zhang. Thermodynamic Modeling and Assessment of Ionic Liquid-based CO_2 Capture Processes, ***Ind. Eng. Chem. Res.***, 2014, 53 (29): 11805

429594 Volumetric, Transport Properties and Microstructures for Binary System of 1-Butyl-3-Methylimidazolium Tetrafluoroborate ([C_4mim][BF_4]) with Water in the Temperature Range (283.15 to 373.15) K

Abstract:

Densities and viscosities of binary mixtures containing 1-butyl-3-methylimidazolium tetrafluoroborate with water over the salt-rich region concentrations were measured at a temperature range of (283.15 to 373.15) K. The excess molar volumes were calculated using the measured experimental data and fitted to the Redlich-Kister equation. Excess viscosities were determined as deviations from the log-mixing rule. The results show that the densities and viscosities strongly depend on the water content and temperature. The temperature dependence of density and dynamic viscosity of aqueous mixtures could be described by an empirical, linear relationship and by the Vogel-Fucher-Tammann (VFT) equation, respectively. Importantly, we identified a new correlation between the excess mixture viscosity and the excess molar volume in the ionic liquid rich regime that is consistent with free volume theory. This correlation enables predicting the viscosity of ionic liquids with added water directly from density measurements.

The densities and viscosities data obtained were also discussed in terms of the interactions between cations and anions of the ionic liquid and the water molecules. Ions of ionic liquids are associated with water molecules with the addition of water into the system. The hydration of the ionic liquid results in the reduction of attraction between cations and anions. Small angle neutron scattering (SANS) measurements were used to detect the microstructure changes of the system with varying water content. Structures are revealed in the ionic liquid aqueous solutions, indicating that inverse micelles form from the excess water once the ions are hydrated.

429648 Solubility and Volumetric Behavior of Binary Mixtures of Lipidic Ionic Liquids and Molecular Solutes

Abstract:

Ionic liquids, molten salts with melting points below 100°C, are of interest to chemical engineers because of their vanishingly low vapor pressures and the ability to tune their properties by altering the chemical structures of their constituent cations and anion. Their extremely low volatility has generated great interest in their use as environmentally benign, low emission solvents for separations and reaction processes. Ionic liquids have been designed and developed to have a large variety of chemical functionality and thermophysical properties; however, the vast majority exhibit polar to moderately polar solvent characteristics and are poor solvents for non-polar solutes. As a large number of energy intensive separations processes related to petroleum refining are centered on the separation of mixtures of non-polar species, this is currently a gap in the abilities of conventional ionic liquids.

Our group has developed a class of ionic liquids with considerable non-polar content to promote non-polar-like solvent properties; these properties are introduced by attaching long alkenyl groups to the cation of 1-alkyl-3-methylimidazolium salts. While attaching long, saturated chains results in melting point increases with alkyl chain length (for alkyl chains of about 9 carbons or longer), incorporating a mid-chain cis unsaturation lowers the melting point of the C18 appended salt by 70 °C. This melting point lowering modification was inspired by biologic systems which incorporate such unsaturations to lower the melting point and increase the fluidity of lipid membranes.

In this work, we highlight liquid/liquid and solid/liquid equilibria of binary mixtures of lipidic ionic liquid and molecular species, as well as the volumetric behavior of these solutions. The solubilities of the non-polar solutes are shown to be highly dependent on both the molar volume and shape of the solute with volumetric behavior indicating negative deviations from ideal solution behavior in the dilute to moderately dilute region with a sharp turn towards positive deviations near the solubility limit. Additionally, the ionic liquid is shown to have very low solubility in the non-polar phase, reducing losses if used in extraction processes.

430796 Understanding Water's Disruptive Role on Solubility in Ionic Liquids

Abstract:

A number of ionic liquids (ILs) are currently being sought as solvents of biomass, with many potential advantages over traditional solvents, including low vapor pressure, low flammability and the ability to be tailored. As ILs dissolve biomass however, water that remains trapped within the matrix is released, mixing with the ionic liquid solvent. In small yet sufficient quantities water can significantly disrupt the dissolution process requiring removal of the water from the solvent, which is a costly process. This could potentially be improved through identification of ILs that are more water-tolerant or ways to stifle water's disruptive mechanism, however a molecular understanding of these processes is first required.

Using molecular dynamics we study the ways in which ionic liquids interact with cellulose, a primary component of biomass. We study a set of 15 different alkylimidazolium based ILs with individual strands of cellulose while varying the concentration of water present in the system. We find that without water the anions form hydrogen bonds with cellulose through a complex series of formation and destruction mechanisms that allow these anions to find discrete and stable states. We outline this complex set of pathways and then show how the addition of water to the ionic liquids can significantly alter the transitions from one state to the next. Furthermore we explore water's direct interaction with the ionic liquid that causes it to lose its effectiveness in dissolving cellulose.

440683 Alternative Ammonia Storage Materials for SCR of NO_X

Abstract:

Increase in the stringent diesel emission standards world-wide has led to a clear need for improved emission control system for NO_X. It is well known that ammonia is an active molecule for the selective catalytic reduction of NO_X in the exhaust. Use of liquid ammonia on-board in a pressure vessel has been excluded due to safety issues. Current technology employs the use of Urea dissolved in water or Diesel Exhaust Fluid (DEF) as ammonia source. However, this system has its own limitations such as freezing at low temperature (-12ºC), solid deposit formation in the exhaust line, and difficulties in dosing ammonia at exhaust temperatures below 200ºC. Another major drawback of AdBlue urea solution is its limited ammonia capacity, due to which there has to be an additional automotive fluid distribution system at every gas station. These critical challenges must be overcome for efficient vehicle integration when indirect ammonia storage is applied.

Considering this, a number of alternative materials such as ammonium carbamate and metal amine salts have been proposed for direct injection of ammonia gas. Metal amines are very attractive functional materials due to their remarkable reversible ammonia storage property and a high volumetric ammonia density of up to 93% of that of liquid ammonia. Herein, we have examined a number of ammonium salts and metal amine chlorides as ammonia storage materials. The TGA profiles of several metal ammine chlorides are shown in the figure below. Important properties such as volumetric capacity, gravimetric capacity (moles of ammonia/g) and ammonia release temperature were evaluated. Furthermore, we have synthesized a range of stable mixed metal chloride solid solutions for the ammonia storage. Studying these different materials will help us in developing a technology that enables direct ammonia dosing for NO_X abatement using simple, high-density storage and delivery system.

413464 Self-Template Porous Carbons Derived from Metal-Organic Framework: Synthesis, Characterization and Adsorption Properties

Abstract:

Self-template Porous Carbons Derived from Metal-organic Framework: Synthesis, Characterization and Adsorption Properties

Jun Wang[a], Jiangfeng Yang[a,b] and Shuguang Deng[a]

wangjun@nmsu.edu

[a.] Department of Chemical and Materials Engineering

New Mexico State University

Las Cruces, New Mexico 88003, USA

[b.] Research Institute of Special Chemicals

Taiyuan University of Technology

Taiyuan 030024, Shanxi, P.R. China

Nowadays, energetic need are mainly covered by fossil fuels, and the poorly handling the emissions containing CO_2, CH_4, and NO_x are responsible for the global warming. Among all the potential solutions, physical adsorption is very outstanding due to its low regeneration cost, high selectivity, thermal stability, and high adsorption capacity. Various solid adsorbents have been investigated to meet the increasing requirements of separation and purification technologies for carbon dioxide capture in the contexts of post-combustion (CO_2/N_2) and natural gas upgrading (CO_2/CH_4). Metal-organic frameworks (MOFs) draw lots of attention in this field in past decades, high capacity and selectivity, controllable pore size, and dramatically variability are the brand marks, however, one main drawback is instable to thermal, humidity, and time. Whereas, activated carbon materials are extremely stable to many industry conditions. Thus, we want to develop activated carbon from MOFs while maintaining advantages of MOFs and make it more stable.

In this presentation, we will report novel porous carbon adsorbents derived from MIL-100 $Al_3O(btc)_2(OH)(H_2O)_2{\bullet}X$ (btc = benzenetricarboxylate; X = encapsulated species, i.e., H_2O, etc.) using three different methods, denoted as Al-100 derived activated carbons (AACs). We will define a set of selective parameters to prove AACs are promising adsorbents for industry applications. Multiple characterization methods are applied, such as, thermogravimetric analysis (TGA), X-ray diffraction (XRD), Raman Spectra, scanning electron microscopy (SEM), and transmission electron microscopy (TEM). The highest specific surface area obtained is 1097 $m^2\ g^{-1}$ with 0.82 $cm^3\ g^{-1}$ pore volume. The highest CO_2 capacity obtained is 6.53 and 4.80 mmol g^{-1} at 273 and 298K, respectively. We will also report the high ideal adsorbed solution theory (IAST) selectivity of CO_2/N_2 and CO_2/CH_4 and adsorption breakthrough simulation results

Self-template Porous Carbons Derived from Metal-organic Framework: Synthesis, Characterization and Adsorption Properties

Jun Wang[a], Jiangfeng Yang[a,b] and Shuguang Deng[a]

wangjun@nmsu.edu

[a] Department of Chemical and Materials Engineering

New Mexico State University

Las Cruces, New Mexico 88003, USA

[b] Research Institute of Special Chemicals

Taiyuan University of Technology

Taiyuan 030024, Shanxi, P.R. China

Nowadays, energetic need are mainly covered by fossil fuels, and the poorly handling the emissions containing CO_2, CH_4, and NO_x are responsible for the global warming. Among all the potential solutions, physical adsorption is very outstanding due to its low regeneration cost, high selectivity, thermal stability, and high adsorption capacity. Various solid adsorbents have been investigated to meet the increasing requirements of separation and purification technologies for carbon dioxide capture in the contexts of post-combustion (CO_2/N_2) and natural gas upgrading (CO_2/CH_4). Metal-organic frameworks (MOFs) draw lots of attention in this field in past decades, high capacity and selectivity, controllable pore size, and dramatically variability are the brand marks, however, one main drawback is instable to thermal, humidity, and time. Whereas, activated carbon materials are extremely stable to many industry conditions. Thus, we want to develop activated carbon from MOFs while maintaining advantages of MOFs and make it more stable.

In this presentation, we will report novel porous carbon adsorbents derived from MIL-100 $Al_3O(btc)_2(OH)(H_2O)_2$•X (btc = benzenetricarboxylate; X = encapsulated species, i.e., H_2O, etc.) using three different methods, denoted as Al-100 derived activated carbons (AACs). We will define a set of selective parameters to prove AACs are promising adsorbents for industry applications. Multiple characterization methods are applied, such as, thermogravimetric analysis (TGA), X-ray diffraction (XRD), Raman Spectra, scanning electron microscopy (SEM), and transmission electron microscopy (TEM). The highest specific surface area obtained is 1097 $m^2\ g^{-1}$ with 0.82 $cm^3\ g^{-1}$ pore volume. The highest CO_2 capacity obtained is 6.53 and 4.80 mmol g^{-1} at 273 and 298K, respectively. We will also report the high ideal adsorbed solution theory (IAST) selectivity of CO_2/N_2 and CO_2/CH_4 and adsorption breakthrough simulation results

Self-template Porous Carbons Derived from Metal-organic Framework: Synthesis, Characterization and Adsorption Properties

Jun Wang[a], Jiangfeng Yang[a,b] and Shuguang Deng[a]

wangjun@nmsu.edu

[a] Department of Chemical and Materials Engineering

New Mexico State University

Las Cruces, New Mexico 88003, USA

[b] Research Institute of Special Chemicals

Taiyuan University of Technology

Taiyuan 030024, Shanxi, P.R. China

Nowadays, energetic need are mainly covered by fossil fuels, and the poorly handling the emissions containing CO_2, CH_4, and NO_x are responsible for the global warming. Among all the potential solutions, physical adsorption is very outstanding due to its low regeneration cost, high selectivity, thermal stability, and high adsorption capacity. Various solid adsorbents have been investigated to meet the increasing requirements of separation and purification technologies for carbon dioxide capture in the contexts of post-combustion (CO_2/N_2) and natural gas upgrading (CO_2/CH_4). Metal-organic frameworks (MOFs) draw lots of attention in this field in past decades, high capacity and selectivity, controllable pore size, and dramatically variability are the brand marks, however, one main drawback is instable to thermal, humidity, and time. Whereas, activated carbon materials are extremely stable to many industry conditions. Thus, we want to develop activated carbon from MOFs while maintaining advantages of MOFs and make it more stable.

In this presentation, we will report novel porous carbon adsorbents derived from MIL-100 $Al_3O(btc)_2(OH)(H_2O)_2$•X (btc = benzenetricarboxylate; X = encapsulated species, i.e., H_2O, etc.) using three different methods, denoted as Al-100 derived activated carbons (AACs). We will define a set of selective parameters to prove AACs are promising adsorbents for industry applications. Multiple characterization methods are applied, such as, thermogravimetric analysis (TGA), X-ray diffraction (XRD), Raman Spectra, scanning electron microscopy (SEM), and transmission electron microscopy (TEM). The highest specific surface area obtained is 1097 $m^2\ g^{-1}$ with 0.82 $cm^3\ g^{-1}$ pore volume. The highest CO_2 capacity obtained is 6.53 and 4.80 mmol g^{-1} at 273 and 298K, respectively. We will also report the high ideal adsorbed solution theory (IAST) selectivity of CO_2/N_2 and CO_2/CH_4 and adsorption breakthrough simulation results.

437237 Nanocomposite Particle Synthesis Using Switchable Ionic Liquids (SWILs)

Abstract:

Anisotropic nanomaterials with controlled morphology play a critical role in product manufacturing incorporating nanotechnology and in modern material chemistry. Anisotropic nanomaterials such as nanowires and nanorods often exhibit interesting properties that are critical for designing devices with desired functions, but their controlled synthesis still remains a grand challenge. Currently available solution-based processes including traditional ionic liquids are inefficient (poor control over the surface morphology, physical and chemical properties), use expensive heterogeneous templates, harsh reaction conditions, hence prohibitive. To exploit the potential of these materials for various applications, a controlled synthesis approach is highly desired. Recently, we developed a novel green synthetic strategy using switchable ionic liquids (SWILs) for the synthesis of nanomaterials. SWILs are mixtures of an alcohol and an amine base with an acidic trigger gas (CO_2, CS_2, COS or SO_2). Ease of SWIL synthesis and excellent control to turn on or off the ionic character of the solvent makes them ideal solvents for the synthesis of nanomaterials. Switch enable us the ability to dissolve high concentrations of precursors in a polar environment (SWILs) and then promote nucleation, growth, and isolation of nanoparticles (Figure 1). Synthesis of carbon rods with 20-100 nm thickness and 1-10 micron length using switchable ionic liquids (SWIL) and its unexpected humidity-driven water expulsion behavior will be discussed. The effect of chemical trigger (CO_2 *vs* CS_2) used in the preparation of SWIL on the formation of nanomaterials will also be discussed.

431312 Pilot-Scale Multistage Fluidized Bed Adsorber

Abstract:

Fluidized bed adsorbers offer several advantages when treating large gas flow rates, including low pressure-drop and high mass and heat transfer. This research systematically investigates the effect of design parameters on a multistage fluidized bed adsorber's efficiency for capturing volatile organic compounds (VOC). Specifically, the effects of number of stages, weir height, and solid/gas ratio on the adsorber removal efficiency were investigated using 1,2,4-trimethylbenzene (TMB) as a test adsorbate and beaded activated carbon (BAC) as adsorbent. The BAC and the TMB-laden air stream flowed counter-currently. The results indicated a non-linear increase in overall TMB removal efficiency by increasing the number of stages, which is due to using more adsorption capacity of the BAC through the adsorber because of higher overall BAC retention time. However, this is effective only if the BAC overall retention time does not exceed the BAC saturation time inside the adsorber. Additionally increasing the weir height increased the removal efficiency of the adsorber due to increase in BAC retention time on each stage. However, better performance was obtained when increasing the number of stages and keeping the same overall retention time (i.e. increasing the number of stages while decreasing the weir height). Finally, increasing the solid/gas ratio increased the adsorber removal efficiency.

414448 Removal of Toxic Chemicals Using Metal-Organic Frameworks

Abstract:

Metal-organic frameworks (MOFs) have been investigated for a wide range of applications, most notably gas storage, separations, and sensing. Only recently have advances been made in the area of toxic chemical remediation focusing on potentially catalytic chemistries for removal. Here we discuss the use of several metal-organic frameworks to remove chemical warfare agent as well as other less-toxic, yet still problematic, chemicals such as chlorine and alkyl halides. In particular, we focus on multiple removal mechanisms, including hydrolysis as well as nucleophilic attack, to mitigate these chemicals.

438890 Catalytic Ionic Liquids for Biomass Conversion and Upgrading

Abstract:

Owing to its unique ability to dissolve lignocellulosic biomass under relatively mild conditions, 1-ethyl-3-methylimidazolium acetate, [C_2C_1im][OAc], has attracted rapidly growing interest as a solvent in biomass pretreatment. However, the catalysis role of [C_2C_1im][OAc] has been least studied for producing downstream platform chemicals in replacement of petro-derived chemicals. Herein, [C_2C_1im][OAc] is identified as an efficient catalyst for lignin depolymerization and sugar upgrading into platform chemicals under mild conditions. Specifically, [C_2C_1im][OAc] renders efficient lignin depolymerization into phenol derivatives (e.g., guaiacol and eugenol) in the presence of catalysts near 100 °C within 3 h; The carbene generated in the presence of [C_2C_1im][OAc] leads to quantitative self-condensation of 5-hydroxymethylfurfural (HMF) to a high-value biorefinery product, 5,5'-di(hydroxymethyl)furoin (DHMF)-a potential C_{12} kerosene/jet fuel intermediate through benzoin condensation. More catalytic reactions based on [C_2C_1im][OAc] have been proposed.

412702 Sustainable Product Separation from a Silylamine-Type Switchable Electrolyte in Electrochemical Reactions

Abstract:

Electrolyte separation from the products in an electrochemical reaction adds challenges to the overall manufacturing process [1], requiring additional separation units to be installed or causing downstream contamination of the products. While traditional salts serve as high conductivity electrolytes in aqueous systems, their poor solubility in organic solvents restrict electrochemical reaction performance in non-aqueous systems.

The ideal properties of an alternative electrolyte for electrochemical reactions are high conductivity, high solubility for various reactants and efficient separation. Ionic liquids (ILs) can be used as alternative electrolytes. ILs are defined as ionic salts with low melting points. The properties of ILs can be tuned by changing the cation or anion. They can have both good solubility in organic solvents and high conductivity, which leads to more promising organic-based electrochemical systems. In order to resolve both the product separation and electrolyte recovery concerns, switchable IL-like solvents can be used [2].

These switchable systems, known as reversible ionic liquids (RevILs), can be in an ionic state during the electrochemical processing and be changed to a molecular state for product separation and/or its recovery through use of an external stimulus. Silylamines can be used as a type of RevIL. They form a RevIL-state with addition of CO_2 and turn back to a molecular liquid-state with mild heating [3, 4].

We previously found that the conductivity of neat silylamine-type RevILs were insufficient to be used as an electrolyte when the model silylamine (3-aminopropyl) triethoxysilane (TEtoxySA) was examined electrochemically [5]. However, the RevIL conductivity drastically increased up to electrochemically-appropriate values in the presence of protic organic solvents [5], though it did not increase with aprotic organic solvents near as much.

Now that the concept of a switchable electrolyte has been proposed where solutions of TEtoxySA-RevIL and protic solvents have high conductivity, high solubility in organic solvents and a simple separation procedure with mild heat, they will be investigated in electrochemical applications. In this presentation, we report the electrochemical reaction and separation procedure using these new switchable electrolytes. Also, the effect of the switchable electrolytes on the reaction will be presented, comparing the system with traditional aqueous electrolytes. 1 Meeting)

References

[1] A.J. Fry, Electrochemical processing, organic, Kirk-Othmer Encyclopedia of Chemical Technology, John Wiley & Sons, Inc.2000, pp. 652-685.

[2] P.G. Jessop, S.M. Mercer, D.J. Heldebrant, CO_2-triggered switchable solvents, surfactants, and other materials, Energy & Environmental Science, 5 (2012) 7240-7253.

[3] A.L. Rohan, J.R. Switzer, K.M. Flack, R.J. Hart, S. Sivaswamy, E.J. Biddinger, M. Talreja, M. Verma, S. Faltermeier, P.T. Nielsen, P. Pollet, G.F. Schuette, C.A. Eckert, C.L. Liotta, The synthesis and the chemical and physical properties of non-aqueous silylamine solvents for carbon dioxide capture, ChemSusChem, 5 (2012) 2181-2187.

[4] J.R. Switzer, A.L. Ethier, E.C. Hart, K.M. Flack, A.C. Rumple, J.C. Donaldson, A.T. Bembry, O.M. Scott, E.J. Biddinger, M. Talreja, M.-G. Song, P. Pollet, C.A. Eckert, C.L. Liotta, Design, synthesis, and evaluation of nonaqueous silylamines for efficient CO_2 capture, ChemSusChem, 7 (2014) 299-307.

[5] J.D. Jimenez, S. Jung, E.J. Biddinger, Ionicity analysis of silylamine-type reversible ionic liquids as a model switchable electrolyte, Journal of The Electrochemical Society, 162 (2015) H460-H465.

418542 Precipitation of Cellulosic Biomass from Ionic Liquid/Cosolvent Mixtures By a Non-Reactive and Energy Efficient Gas Antisolvent Method

Abstract:

Select ionic liquids (ILs) demonstrate inherently large cellulose dissolution capacities making them highly touted solvents for biomass processing. Through an ionic liquid based dissolution and precipitation process cellulose crystallinity is largely reduced consequently enhancing subsequent chemical and biochemical reaction processes. Conventional antisolvents for biomass precipitation from ILs include polar protic liquids (i.e. H_2O, EtOH, etc.) which are highly effective at disrupting IL-cellulose interactions. Recently we have found that these liquid antisolvents are so effective that as little as 1 mass% residual water on an IL can reduce its cellulose capacity by as much as 55%. Therefore, ILs must be highly purified from liquid antisolvents prior to recycle. Preliminary analyses indicate that quantitative separation of an IL from a liquid antisolvent is highly energy intensive and could potentially impede large scale viability. Therefore, as an alternative to conventional liquid antisolvent processes we have developed a novel gas antisolvent method which precipitates cellulosic biomass from IL/cosolvent mixtures by compressed carbon dioxide at low to moderate pressures (i.e. 23 to 66 bar). The gas antisolvent separation process is especially unique as it is non-reactive and completely reversible. By simple depressurization of CO_2 to just a few bar pressure below the separation point, ionic liquid solvation power for cellulose is completely regenerated. This presentation will highlight the solid-liquid phase equilibrium effects of both conventional and novel separation processes. Additionally, the advantages of polar aprotic cosolvents will be highlighted as they apply to cellulose dissolution in IL/solvent mixtures and the gas antisolvent process. Spectroscopic techniques will identify key trends within the separation data. Finally an energy analysis will be presented to demonstrate the advantages of this novel CO_2 relative to liquid antisolvent separations.

419589 Nanoengineered Solid Acid Catalysts for Green Processes

Abstract:

Nanoengineered Solid Acid Catalysts for Green Processes

H. Liu, P.Shen, H.T. Zhang, S.J. Zhang*

Beijing Key Laboratory of Ionic Liquids Clean Process, Key Laboratory of Green Process and Engineering, Institute of Process Engineering, Chinese Academy of Sciences, Beijing, 100190, P. R. China

E-mail: htzhang@ipe.ac.cn

Improving catalytic activity and selectivity as well as reusability of catalysts is an unavoidable challenge because of stringent economical and ecological desire for sustainability on account of the principle of "green chemistry" and "green technology". The development of nanosciences and nanotechnologies make it possible to fabricate complicated artificial nanoarchitectures *via*the elf-assemble and reorganization of different anisotropic nanoscale building blocks (nanocrystals, nanosheets, nanowires, and polyhedrons of oxides, metal, sulfide, and nitrite).

Among different kinds of acid catalysts, solid acid catalysts are attracting more attention owing to their obviously environmental benign characteristics with respect to weak corrosiveness, safety, high efficiency, easiness of separation and recovery. Solid acid nanosheets are promising because of their exotic physicochemical properties and high specific surface areas and tunable acidity. They represent a diverse and largely untapped source of basic building blocks for rational creating complicated nanocatalysts with designable microstructures and programmed propertiesthe. In addition, many nanoparticles could be employ as basic building blocks with nanosheets to induce some synergistic effects that could improve catalytic performance. Our research study indicated that nanoengineered solid acid catalysts exhibited improved catalytic performance for Friedel-Crafts alkylation, acetalization and photocatalytic reduction reactions. Here we mainly present some progresses in retrosynthesizing complication layered nanohybrids by using solid acid nanosheets, and understanding the relationships between microstructures and improved catalytic performance.

427326 Synergistic Catalytic Performance in the Synthesis of Cyclic Carbonates Based on Ionic Liquids

Wednesday, November 11, 2015
Exhibit Hall 1 (Salt Palace Convention Center)
Suojiang Zhang, Institute of Process Engineering, Chinese Academy of Sciences, Beijing, China

Synergistic Catalytic Effect in the Synthesis of Cyclic Carbonates Based on Ionic Liquids

Suo-Jiang Zhang , Wei-Guo Cheng

Key Laboratory of Green Process and Engineering, Institute of Process Engineering, Chinese Academy of Sciences, P.O. Box 353, Beijing, 100190, China

**Corresponding author: sjzhang@home.ipe.ac.cn*

Carbon dioxide released annually to the atmosphere at ca. 30 000 Mt1 is also an attractive C1 feedstock as it is nontoxic, abundant, inexpensive. From the viewpoint of environmental protection and resource utilization, it is important to transform CO_2 into cyclic carbonates efficiently not only in academia but also in industrial applications(Scheme 1). Many scientists have made a concentrated effort to improve the performance of the ionic liquids (ILs) catalysts during the past decade due to their tunable structures. Synergistic catalytic effects of ILs catalytic system play an important role of promoting the cycloaddition reactions of epoxides. However, due to the diversity of different anion¨Ccation combinations it is difficult to make generalizations about synergistic effects.

THE GRAPHIC ON THIS PAGE WAS NOT AVAILABLE ON THE E-MEDIA VERSION OF THIS PAPER.

Scheme 1 Fixation CO_2 and epoxides into cyclic carbonates

In this work, we have investigated the synergistic effect in the catalytic system based on ILs through experimental and theoretical studies[1-5]. The results show that metal and ionic liquids, cations and anions of ionic liquids, the functional groups in cations or supports all have a synergistic effect on the cycloaddition reactions. Synergistic catalytic effects of ILs play an important role of remarkably reducing the activation energy and promoting ring-opening of epoxide. The design of catalytic system of ionic liquids can be possible as the synergistic effect on a molecular level is understood by using density functional theory (DFT) studies and in situ characterization. Significant developments in IL catalysts design and optimal process development for the synthesis of cyclic carbonates have been made with the possible prediction of the synergetic effect based on ILs.

Keywords: synergistic effect, ionic liquid, carbon dioxide, cyclic carbonate, process

References

1. Zhang SJ, Sun J, Zhang XC, Xin JY, Miao QQ, Wang JJ. Ionic liquid-based green processes for energy production. *Chemical Society Reviews*, 2014, 43: 7838-7869

2. Cheng WG, Chen X, Sun J, Wang JQ, Zhang SJ. SBA-15 supported triazolium-based ionic liquids as highly efficient and recyclable catalysts for fixation of CO_2 with epoxides. *Catalysis Today*, 2013,200: 117- 124

3. Sun J, Wang JQ, Cheng WG, Zhang JX, Li XH, Zhang SJ, She YB. Chitosan functionalized ionic liquid as a recyclable biopolymer-supported catalyst for cycloaddition of CO_2. *Green Chemistry*, 2012, 14: 654-660

4. Wang JQ, Sun J, Cheng WG, Dong K, Zhang XP, Zhang SJ. Experimental and theoretical studies on hydrogen bond-promoted fixation of carbon dioxide and epoxides in cyclic carbonates. *Phys. Chem. Chem. Phys.* 2012, 14: 11021- 11026.

5. Xu BH, Wang JQ, Sun J, Huang Y, Zhang JP, Zhang SJ. Fixation of CO_2 into cyclic carbonates catalyzed by ionic liquids: a multi-scale approach. *Green Chemistry*, 2015, 17: 108-122.

429678 Kinetics of Hydroformylation of Ethylene Using Homogeneous Rhodium-Phosphine Complexes

Abstract:

The hydroformylation reaction is a useful method to functionalize C-C bonds for providing aldehyde and alcohol compounds, and is also an important industrial catalytic process. In view of the importance, it is attracting continuous and increasing academic and industrial interest. Studies on the kinetics of hydroformylation of olefins not only are instructive for improvement of the catalytic system but also provide the basic information for design and scale-up of appropriate reactors.

In this research, the kinetics of hydroformylation of ethylene with triphenylphosphine modified rhodium catalyst was studied. The reaction was carried out batchwise in a flat-bottomed, magnetically stirred 150 mL steel autoclave equipped with a 120 mL teflon tube. Under the conditions of the Rh concentration of 0.0015mol/L, PPh_3/Rh=10, C_2H_4: CO: H_2 =1: 1: 1 (molar ratio), 2 Mpa and 100 °C, the selectivity for propionaldehyde was nearly 100%. The ethylene hydroformylation kinetics was studied as the function of the amount of catalyst and triphenylphosphine, partial pressure of ethylene, CO and H_2, and temperature. All these parameters were found to influence the rate of hydroformylation. On the range of conditions (catalyst, 1.024~2.2658 mmol·L^{-1}; PPh_3, 0.0058~0.3018 mmol·L^{-1}; P_{H2}, 0.26~1.27 MPa; P_{CO}, 0.11~1.22 MPa; P_{C2H4}, 0.18~1.13 MPa; T, 355.15~372.15 K), rates were observed to increase with the increase in the amount of the catalyst and triphenylphosphine; rates increased on increasing the C_2H_4 or H_2 pressure; rates increased on increasing the CO pressure up to certain values, and then decreased. The activation energy was found to be 61.52 kJ·mol^{-1}. A kinetic rate model based on the mechanism of hydroformylation of ethylene was found to fit with the experimental rate with ±5% deviation.

429744 Dopant Effect on Perovskite Based Mixed Metal Oxides for Chemical Looping with Oxygen Uncoupling (CLOU)

Abstract:

As an alternative to conventional combustion, chemical looping combustion (CLC) has the potential to achieve higher 2nd law efficiency with inherent CO_2 capture capabilities. Although CLC has been extensively demonstrated for gaseous fuels, extension of this process to solid fuels, such as coal, is challenging due to the slow kinetics of solid-solid reactions. Chemical Looping with Oxygen Uncoupling (CLOU) provides a possible solution to this challenge. In a CLOU process, lattice oxygen from the oxygen carrier is released into the gas phase for facile combustion of solid fuels. The oxygen deprived oxygen carrier is then regenerated with air while producing heat. In the current study, perovskite-structured oxygen carriers with a general formula of $A_xA'_{1-x}MnO_3$ (A/A' = Sr, Ca, Ba) are investigated for CLOU applications. Results show $CaMnO_3$ provides excellent oxygen uncoupling ability; however, irreversible phase decomposition is observed. The effects of Sr and Ba as A-site dopants on $CaMnO_3$ are explored with the primary intention to prevent irreversible phase decomposition. These materials are tested for oxygen release properties, redox stability, and reactivity. It was determined that both barium and strontium doping lead to decreased oxygen release at high temperatures. Furthermore, Sr dopant exhibits excellent compatibility with the $CaMnO_3$ parent structure at a wide range of concentrations. $Ca_{1-x}Sr_xMnO_3$ is also able to reduce the initial oxygen release temperature by over 400°C, enabling facile oxygen donation. With the goal of understanding the effect of A-site cation in perovskite based oxygen carriers, DFT calculations were performed to determine the vacancy formation energy for these perovskite materials. DFT results indicate CaMnO
lattice oxygen in the perovskite structure. The finding is used to explain CLOU properties of the aforementioned oxygen carriers and to guide future efforts to improve oxygen carriers.

429567 Influence of Percolation Threshold on the Redox Performance of CeO2-Supported Cu-Based Materials

Abstract:

A mid-term solution to stabilize the concentration of CO_2 in the atmosphere is the implementation of CO_2 capture and storage technologies (CCS). Chemical looping combustion (CLC) is a promising CCS technology and conceptually linked to oxy-fuel combustion. In CLC, the combustion of a hydrocarbon takes place in the presence of an oxygen carrier, typically a transition metal oxide. In CLC, the oxygen required for the combustion of a hydrocarbon comes from an oxygen carrier, *viz.*

$$C_nH_{2m}\ (g) + (2n + m)Me_xO_y\ (s) \rightarrow nCO_2\ (g) + mH_2O\ (g) + (2n+m)Me_xO_{y-1}\ (s)$$

Desorption of the reaction product(s) generates an oxygen vacancy on the surface which is filled subsequently by oxygen from the bulk. Therefore, the CLC performance depends critically on the activation energy for transport of ions and electrons through the oxygen carrier. Cu is a very attractive material for CLC owing to its (i) high oxygen carrying capacity of 0.25 g O_2/g Cu and (ii) low tendency for carbon deposition. However, Cu has one main disadvantage, *i.e.* a low Tammann temperature of 405 °C and, thus, requires stabilization by a high Tammann temperature support *e.g.* CeO_2. The advantage of using CeO_2 as a support over other conventional ceramics (*e.g.* $MgAl_2O_4$) is that CeO_2-based cermets can contribute to the storage and release of oxygen *via* the Ce^{4+} - Ce^{3+} transition. Here, we investigate the influence of the percolation threshold on the redox kinetics of CeO_2-supported Cu. We used a co-precipitation technique to synthesize CeO_2-stabilized, Cu-based materials containing 20 - 60 wt. % Cu. The synthesized oxygen carriers were characterized using 4-point conductivity measurements, energy dispersive X-ray (EDX) spectroscopy, temperature programmed reduction (TPR) and scanning electron microscopy (SEM). The cyclic redox stability of the synthesized materials was evaluated in a thermo-gravimetric analyzer (TGA) at 900 °C using 10 vol. % H_2 in N_2 as the fuel and air for re-oxidation.

For Cu-rich materials, EDX mapping revealed the presence of CeO_2 islands in a continuous Cu network, therefore, the transport of oxygen ions and electrons predominately occurred *via* Cu/CuO conduction pathways. Below the percolation threshold of Cu, CeO_2 formed a continuous network containing CuO islands. The conductivity measurements showed that in this regime the conduction of electrons and ions occurred through CeO_2 grains. Moreover, conductivity and performance measurements showed that the activation energy for charge transport through the material increased with decreasing Cu content, which in turn decreased the rate of oxidation of the material. Owing to the low activation energy for charge transport and fast rate of oxidation, the material containing 60 wt. % Cu was identified as a promising material for CLC.

428578 Fe2O3/Ce-Î³Al2O3 As an Oxygen Carrier for Liquid Fuel Based Clc

Abstract:

The development of a highly reactive, stable, economic and environmentally friendly oxygen carrier is considered the key issue in the large scale application of chemical looping combustion. The availability of Iron, besides being cheap, nontoxic and environmentally friendly made it commercially attractive for CLC application. The main objective of this work is to improve the reactivity of Fe.

The oxygen carriers were synthesized using incipient wetness technique [3]. The starting materials are γ-Al_2O_3, $Ce(NO_3)_3.6H_2O$ and $Fe(NO_3)_3.9H_2O$. The preparation of the oxygen carrier consists of four major steps: impregnation, drying, reduction and finally calcination. First the support was modified using 1 wt% Ce before loading the metal in 4 cycles, 5 wt% each to reach a final concentration of 20 wt%.

The oxygen carrier reactivity and its ability to withstand the repeated oxidation/reduction cycles are investigated through successive TPO/TPR experiments. The samples reducibility and oxygen carrying capacity can be also extracted from TPR profiles. The reduction profile of the oxygen carrier shows a slightly narrow peak at around 357 °C and a wider peak at 564 °C. The first reduction peak corresponds to the reduction of Fe_2O_3 to Fe_3O_4 while the second corresponds to the transition of Fe_3O_4 to FeO to Fe. The reduction percentage of the oxygen carrier is around 88%, which remained stable through 10 TPR/TPO cycles.

418016 Influence of Supports on Chemical Looping Methane Reforming Activity of NiO-Based Oxygen Transfer Materials

Abstract:

Influence of supports on chemical looping methane reforming activity of NiO-based oxygen transfer materials

Introduction

Hydrogen is an energy carrier that can be used to efficiently store regenerative energy, and it fulfills the requirements of being the most environmentally friendly fuel for several applications. Although the steam methane reforming process is currently the major source of hydrogen production, its high energy demands along with increased CO_2 production make the overall process undesirable both from economic and environmental considerations. The Sorption Enhanced Chemical Looping Steam Methane Reforming (SE-CL-SMR) is an emerging technology that can simultaneously minimize the aforementioned drawbacks and produce pure hydrogen. The stable and active oxygen transfer materials (OTMs) and CO_2 sorbents are the major requirements for a viable SE-CL-SMR process. NiO supported OTMs have been promising oxygen carriers for chemical looping methane reforming reactions due to their superior redox reactivity, catalytic performance, stability and negligible volatility. In this study we investigated the influence of supports on redox activity of OTMs in the absence of steam in a thermo-gravimetric analyzer (TGA) unit, and in conventional and chemical looping steam methane reforming reactions (CLR) in a fixed bed flow unit.

Experimental

Four NiO-based OTMs with 40wt% NiO loading were synthesized by wet impregnation using commercial supports:

(SiO_2, TiO_2, Al_2O_3, ZrO_2). Both pre- and post- characterizations of the OTMs such as BET, XRD, H_2-TPR, TPO and TEM were performed. Twenty redox cycles, methane reduction at 650°C/air oxidation at 800°C, were conducted in a TGA unit connected to a mass spectrometer (MS). Conventional steam reforming activity of the pre-reduced OTMs was tested at 650°C in a bench scale fixed bed flow unit with a steam/methane (S/C) ratio of 3 and GHSV=10^5h^{-1}. In a second step, NiO/ZrO_2 and NiO/Al_2O_3, which exhibited satisfactory reforming activity and stability, were tested under CLR conditions. The materials were initially exposed to CH_4/steam in their oxidized state at 650°C for 1h (S/C=3). At the end of reforming stage, metallic Ni was re-oxidized with air at 850°C for 15min and the cycle was then repeated.

Results and discussion

BET surface area of the 40%NiO-OTMs decreased after loading NiO compared to the pure supports. XRD confirmed the presence of NiO phase in all OTMs, and strong metal-support interaction compounds ($NiAl_2O_4$ and $NiTiO_3$) were identified in Al_2O_3- and TiO_2- supported OTMs. During the 20 redox cycles in TGA, all OTMs except 40%NiO/g-Al_2O_3 attained very high reduction degrees (94% - 100%) (Fig. 1a). Degree of reduction (RD, %) of NiO/Al_2O_3 slowly increased up to cycle 11 reaching a maximum and steady value of 86%. This relatively low reducibility is mainly attributed to the strong metal-support interaction, as identified by XRD. In order to investigate the unusual trend in the reduction degree of NiO/Al_2O_3, temperature programed reduction (TPR) was performed for fresh and used (after 10th redox cycles) OTM in the TGA (not shown). Results confirmed the decomposition of $NiAl_2O_4$ and increase in the $NiO/NiAl_2O_4$ ratio with cycles, indicating that re-crystallization of metallic Ni after reduction in methane weakens the alumina-nickel oxide interaction. All four OTMs underwent complete re-oxidation during the oxidation step of the cycling experiments, reaching the same oxidation degree level of 100% in all cycles. The H_2, CO, CO_2 and H_2O were the products detected during CH_4 reduction, whereas CO_2 was the only oxidation product detected by MS.

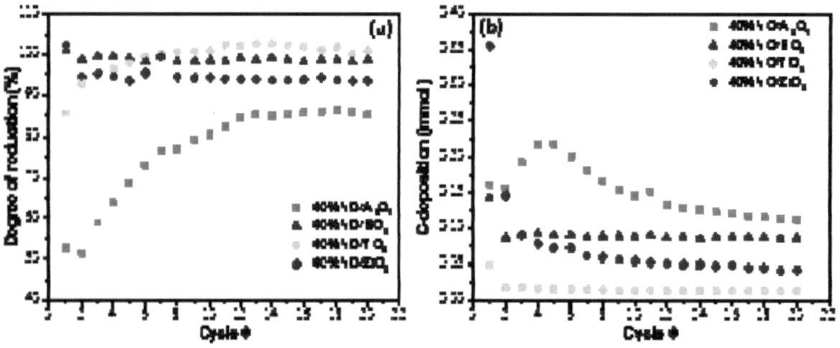

Figure 1 The variation in (a) the reduction degree and (b) the amount of carbon with cycles

The amount of carbon deposited during reduction steps, calculated from the observed weight gain in TGA, is shown in Fig. 1b. NiO/Al_2O_3 OTM exhibited a volcano shape curve which is possibly caused by changes in catalyst structure (gradual $NiAl_2O_4$ reduction) with cycles (Fig. 1b). For all OTMs trends in hydrogen formation (not shown) followed the same trend as C-deposition, indicating that there is a relationship between deposited carbon and H_2 formation during the reduction step. Data analysis showed that hydrogen production occurs primarily via CH_4 split. The amount of carbon deposition and hydrogen production was higher during initial cycles on Al_2O_3 and ZrO_2 OTMs, followed by decrease in subsequent cycles. This indicates deactivation of these OTMs towards the CH_4 decomposition reaction.

In order to investigate the nature and types of carbon deposited on OTMs by methane decomposition during the reduction cycles, TPO and TEM analysis were performed. TPO characterization of all OTMs after the first methane reduction step in TGA showed a single CO_2 peak centered at ~500 C – 550 C indicating the presence of one type of carbon on the surface. However the TPO profile of NiO/Al_2O_3 after the methane reduction step of the fourth cycle indicated the presence of two carbon species, oxidized to CO_2 at 575 C and 700 C respectively. In order to obtain further insights into the nature of the deposited carbon species, the samples were examined with transmission electron microscopy. Both graphitic and filamentous carbon were observed on Al_2O_3 and ZrO_2 supported OTMs. TEM image of Al_2O_3 supported OTM after the first cycle is shown in Fig. 2 for illustration. On the other hand, only graphitic carbon was observed on the SiO_2 and TiO_2 supported OTMs.

Experiments under conventional reforming conditions (with pre-reduced OTMs) showed a satisfactory performance of NiO/Al_2O_3 and NiO/ZrO_2 at 650°C, with a high methane conversion and less than 8% deactivation after 10 h on stream. On the other hand, NiO/SiO_2 and NiO/TiO_2 were found to be inactive, with initial CH_4 conversion less than 12%. The two most promising materials were then tested under chemical looping reforming conditions for 20 consecutive steam methane reforming/air oxidation cycles. Methane conversion as a function of cycles is shown in Figure 3. NiO/ZrO_2 exhibited a satisfactory activity with initial CH_4 conversion around 80% and less than 2% deactivation after 20 cycles, corresponding to 20h exposure to CH_4/steam. Initial CH_4 conversion on NiO/Al_2O_3 OTM was similar (~79%), but it deactivated more rapidly, leading to a CH_4 conversion of 59% in the 20th cycle (Fig. 3). This loss of activity can be attributed to the lower hydrothermal stability of alumina compared to zirconia under the steam reforming conditions.

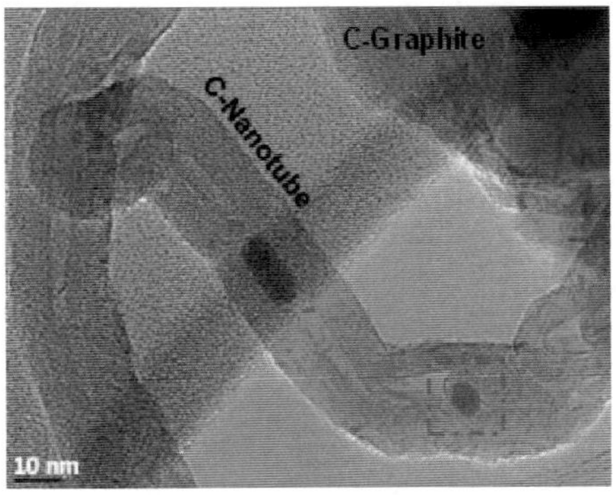

Figure 2 TEM image of Al_2O_3 OTM after 1st reduction cycle

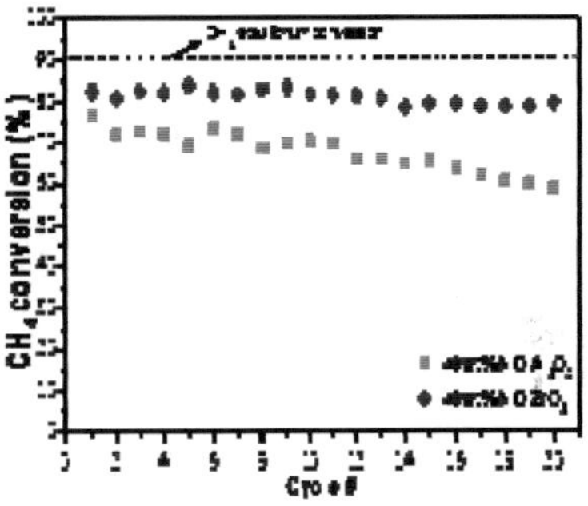

Figure 3 CH_4 conversion in chemical looping reforming experiments for NiO/ZrO_2 and NiO/Al_2O_3 *(Reduction-Reforming: 650 C, 1h, S/C = 3, GHSV =100000h-1; Oxidation: 800 C, 15min, Air)*

Conclusions

Support was found to have a significant effect on performance of NiO oxygen carriers in chemical looping steam methane reforming. All investigated OTMs supported on silica, alumina, zirconia and titania displayed satisfactory activity and high stability during 20 methane reduction/air oxidation cycles in TGA experiments. The 40%NiO/Al_2O_3 OTM showed unusual behavior compared to the other OTMs due to gradual reduction of metal-support compounds to more easily reducible species with cycles. Carbon deposition in the absence of steam in the TGA occurs primarily via CH_4 decomposition on reduced Ni. Only graphitic carbon was observed on SiO_2- and TiO_2- OTMs, whereas both graphitic and filamentous carbon were observed on Al_2O_3- and ZrO_2- OTMs. Testing of all OTMs as reforming catalysts in conventional SMR showed that NiO on titania and silica is inactive, while alumina and zirconia supported OTMs had high activity and stability characteristics. The latter two materials were tested in the chemical looping methane reforming process, where they exhibited promising performance. NiO/ZrO_2 showed good activity with initial CH_4 conversion around 80% and less than 2% deactivation after 20 cycles. NiO/Al_2O_3 had also a high initial CH_4 conversion, but had higher deactivation rate with cycles, probably due to a lower hydrothermal stability of alumina.

ACKNOWLEDGMENTS

This work was made possible by NPRP grant 5–420–2–166 from QNRF (member of Qatar Foundation). The statements made herein are solely the responsibility of the authors.

418060 Production of Inherently Separated Syngas Streams Via Chemical Looping

Abstract:

Production of Inherently Separated Syngas Streams via Chemical Looping

Amey More and Götz Veser

Department of Chemical Engineering, University of Pittsburgh, Pittsburgh, PA

Abstract:

'Chemical Looping Combustion' (CLC) is an emerging clean combustion technology which offers an efficient and elegant route for fossil fuel combustion with inherent CO_2 capture based on the cyclic oxidation and reduction of an "oxygen carrier" (typically a metal) with air and a fuel, respectively. However, chemical looping can be used for fuel conversion well beyond combustion. We have previously shown that chemical looping allows for highly efficient production of synthesis gas from methane via partial oxidation. In the present work, we propose the formation of separated streams of CO and high-purity H_2 via non-oxidative methane cracking and oxidative removal of the produced carbon using CO_2 in a "looping configuration", i.e. at periodic process conditions. The proposed process utilizes the well-known reactivity of Ni towards C-H bond activation (which also makes Ni/NiO a popular oxygen carrier for conventional chemical looping combustion). CH_4 is cracked catalytically over Ni, producing gaseous H_2 and solid carbon which deactivates the carrier. The carrier is then regenerated by burning off the carbon using CO_2 as oxidizer, enabling the reduction of CO_2 to CO.

Supported Ni carriers were synthesized using a simple wet-impregnation procedure and characterized using X-Ray Diffraction (XRD) and Electron Microscopy (SEM, TEM, EDX). Solid state and gas phase conversions and selectivities were determined via thermo-gravimetric analysis (TGA) combined with fixed bed reactor studies. We found that the nature of the support strongly influences particle size and distribution of the active metal species (Ni) and thus impacts the overall activity of the carrier towards methane cracking. Furthermore, these differences also impact the morphology of the carbon formed during the cracking reaction, which further impact overall carrier reactivity. Similarly, the oxidative removal of the carbon by CO_2 depends strongly on the type of carbon formed and hence on support properties. Based on these observations, suitable process conditions were identified and a simple, yet promising carrier system for an efficient chemical looping process for the production of separated syngas streams (i.e. separate H_2 and CO effluent streams) is proposed.

434296 Novel Thermochemical Synthesis of Ammonia and Syngas from Natural Gas

Abstract:

Natural gas is connected to ammonia and syngas production via a novel thermochemical approach. Ammonia is required as a main constituent of fertilizer and is widely recognized as at least partially responsible for the boom in food production known as the "green revolution" of the early twentieth century. However global food production will need to double due to expected increase in world population to 9.6 billion by 2050 and rising demand for protein among developing nations. Currently global ammonia production is greater than 140 million metric tons annually and will need to increase to keep up with the expected increase in fertilizer demand for food production. The conventional Haber-Bosch process for ammonia synthesis operates at temperatures as high as 500°C and pressures reaching 300 atmospheres. Per ton of ammonia produced, the Haber-Bosch process uses 0.6 tons of natural gas as both a source of hydrogen and combustion energy and emits almost 1.6 tons of carbon dioxide.

In the present work we report an experimental proof-of-concept of a thermochemical reaction cycle to synthesize ammonia and syngas at near-ambient pressure from nitrogen, water, and methane. Manganese nitride ($Mn_6N_{2.58}$), synthesized from its elements, is reacted with steam to produce ammonia and manganese oxide (MnO). Results show 43-50% of the lattice nitrogen in the nitride mixture is converted to ammonia while all manganese is converted to the oxide. MnO was reacted with a dilute methane in nitrogen stream yielding the manganese nitride, synthesis gas, and some solid carbon. The presence of carbon dioxide is not observed.

Based on these results, employment of a thermochemical cycle appears a promising method to produce ammonia from atmospheric nitrogen, steam, and natural gas at near-ambient pressures. Significantly reduced carbon dioxide emissions while producing a valuable product in synthesis gas present a key advantage for this novel thermochemical approach going forward as global ammonia production increases to enable food production to double by 2050.

425785 Chemical-Looping with Oxygen Uncoupling (CLOU): A Process Development Perspective

Abstract:

Chemical-looping with oxygen uncoupling (CLOU) has emerged to be one of the promising options for combustion of carbonaceous solid fuels, offering the promise of a reduced energy penalty for capture of carbon dioxide. The potential of the CLOU process for solid fuel combustion with copper oxide as an oxygen carrier has been demonstrated through process models and energy integration studies for steam production[1, 2].

Since the first reference to the use of CuO as an oxygen carrier by Lewis, Gilliland and Sweeney in 1951[3] followed by the seminal experiments on Chemical -looping with oxygen uncoupling (CLOU) at Chalmers in the later part of the last decade[4], there has been considerable research on the topic. The research on CLOU covers a broad spectrum of topics, e.g. combustion fundamentals, oxygen carrier development, gas-solid reaction kinetics, fluidized beds, process engineering analysis. The talk will highlight upon the uniqueness of CLOU, and present a comparison with some of the established fluidized bed technologies (e.g. Fluid Catalytic Cracking regenerators and circulating fluidized bed coal combustors) from the perspective of process engineering considerations for deriving design insights.

[1] Sahir, A.H. et al., A comparative process study of chemical-looping combustion (CLC) and chemical-looping with oxygen uncoupling (CLOU) for solid fuels, International Journal of Greenhouse Gas Control, 2014, 22, 237-243.

[2] Dansie, J.K. et al. , An investigation of steam production in chemical-looping combustion (CLC) and chemical-looping with oxygen uncoupling (CLOU) for solid fuels, Chemical Engineering Research and Design, 2015, 94, 12-17.

[3] Lewis W.K. et al. ,Gasification of Carbon Metal Oxides in a Fluidized Powder Bed, Chemical Engineering Progress, 1951,47, 5, 251-256.

[4] Mattisson T. et al. ,Chemical-looping with oxygen uncoupling for combustion of solid fuels, International Journal of Greenhouse Gas Control, 2009,3, 11-19.

**Current Affiliation of Asad H. Sahir: Biorefinery Analysis and Exploratory Research, National Bioenergy Center, National Renewable Energy Laboratory, Golden, CO, 80401. This presentation is based on Asad's work at the University of Utah

431019 Autothermic Chemical Looping Process for Hydrogen Production Using Activated Carbon As a Catalyst

Abstract:

Methane thermo-catalytic decomposition has been recognized as an advanced process for hydrogen (H_2) production/natural gas decarburization in the carbon constrained world due to the low CO_2 formation with respect to energy supply. As an alternative to costly transition metal catalysts, it is believed that carbon materials are promising catalysts for this process, of which activated carbon (AC) provides reasonable reaction kinetics.

The application of a chemical looping process to methane thermo-catalytic decomposition using AC as a catalyst for continuous high purity H_2 production has been studied. This process includes two interconnected reactors, a decomposer and a regenerator. In the decomposer, with the aid of AC catalyst, methane is continuously converted to solid carbon and H_2 via methane thermo-catalytic decomposition, and a nearly pure H_2stream can be produced under appropriate conditions. The spent AC catalyst which loses the activity to some extent, is regenerated in the regenerator for looping operation. Two major technical barriers need to be overcome in the regenerator to implement this process. First, the catalytic activity of the deactivated AC catalyst should be completely recovered, and second, the regeneration process should provide enough heat to support the endothermic methane decomposition reaction in the decomposer. Heat can be transferred from the regenerator to the decomposer via the AC catalyst in the form of sensible heat.

The regeneration agents, such as O_2 and steam, play a detrimental role. O_2 regenerated AC catalyst can provide enough heat, however, the activity can't be completely recovered. AC catalyst regenerated by steam can completely recover the activity, but requires an extra heat for the endothermic gasification. The use of steam with O_2 as regeneration agents was studied on TGA apparatus and a fixed bed reactor, and the results show that both requirements can be satisfied under certain mixing ratios. The product syngas from the regenerator has a H_2 to CO ratio of about 0.7. The addition of CH_4 to the regeneration agents to further increase the H_2 to CO ratio to greater than 2 by methane partial oxidation is also investigated, hence the product syngas can be directly used for chemical production. Heat balance analysis and catalytic activity comparisons were performed to ensure the two processing barriers were successfully overcome.

The fresh, spent and regenerated AC catalysts were characterized by physical properties (morphology, N_2 adsorption and desorption, mechanical strength) and surface chemistry (oxygen functional groups) studies. Additionally, the evolution of the catalyst with successive decomposition and regeneration cycles was also investigated. Some insights into the performance of the regenerated AC are provided.

ADVANCED SOLIDS FLOW SENSOR TECHNOLOGY FOR CHEMICAL LOOPING

Benjamin T. Chorpening, U. S. Department of Energy, National Energy Technology Laboratory, Morgantown, WV, USA
David W. Greve, Carnegie Mellon University, Pittsburgh, PA, USA
Jared M. Charley, AECOM, National Energy Technology Laboratory, Morgantown, WV, USA
Justin M. Weber, U. S. Department of Energy, National Energy Technology Laboratory, Morgantown, WV, USA
Douglas L. Straub, U. S. Department of Energy, National Energy Technology Laboratory, Morgantown, WV, USA

Short Abstract

Chemical looping combustion is being developed by the U.S. Department of Energy as a possible technology for heat or power generation with integrated carbon dioxide separation. Improved sensors and controls are among the technical challenges that must be overcome for the commercial success of chemical looping. For improved control the chemical looping process, accurate and rapid measurement of key process parameters is required. A high temperature solids-flow sensor, based on microwave Doppler technology, is being developed to provide online measurement of solids circulation rate. This paper reports on progress in the development and testing of this sensor technology for chemical looping combustion conditions.

Introduction

Chemical looping combustion is being developed with support of the U.S. Department of Energy because it has potential for carbon capture with less efficiency penalty than conventional power cycles with post-combustion capture [1,2]. Chemical looping uses an intermediary, typically a metal oxide, to collect oxygen from air in the air reactor. The oxidized particles are then transported to the fuel reactor, where the oxygen carrier particles are reduced. The product gas stream contains mostly carbon dioxide and water vapor, which be readily separated for carbon capture, utilization and storage. An inert buffering gas is injected in the process between the reactors to help separate the oxidation and reduction processes.

One of the challenges for operation of a chemical looping combustion process is the measurement and control of the solids circulation rate. The circulation rate is important for controlling both heat transfer and chemical reaction processes. While the circulation rate may be estimated from pressure drops at various points in the process, this measurement approach is indirect, and has had considerable uncertainty.

While there are many methods employed for measurement of air conveyed solid materials, the operating temperature and pressure parameters used or examined for use in lab-scale or pilot-scale chemical looping systems make most measurement methods for conveyed solids very difficult to implement. Chalmers University (Sweden) [3], and Southeast University (China) [4] built lab-scale reactors using iron at 1000°C and nickel at 1040°C. Proposed chemical-looping gas turbine systems have the highest pressure requirements (up to 21 bar) with temperatures as high as 1050°C for iron-based carriers [5,6]. In contrast, most solids flow measurement technology has been developed for atmospheric pressure, low temperature applications such as conveyance of food ingredients, such as grains and sugar, or raw materials such as plastic pellets.

The Chemical Looping Reactor (CLR) has been designed and constructed at NETL for the purpose of research and development of the chemical looping combustion process. The chemical looping reactor operates at positive pressure at high temperature (~ 1000 °C). The design of a new sensor must take these conditions into account.

Microwave Solids Flow Sensor Development

NETL in collaboration with Carnegie Mellon University, is developing a prototype high temperature microwave Doppler mass flow sensor specifically for the chemical looping application [7,8]. To meet the special requirements of this application, microwave sensing provides several advantages. First, it is a non-contact method which allows a pressure boundary. Second, it is less sensitive to the surface roughness of the pressure boundary, or microwave window, than an optical (visible wavelength) method. Third, it has previously been commercially implemented for slightly elevated temperatures (up to 200°C). The project approach taken allows modification of the antenna module design, electronics, and data processing to fit the special application requirements. The electronics and a first prototype microwave antenna module (Figure 1) have been tested at room temperature, and data processing has been developed [9]. The prototype system has been included in testing during chemical looping operations.

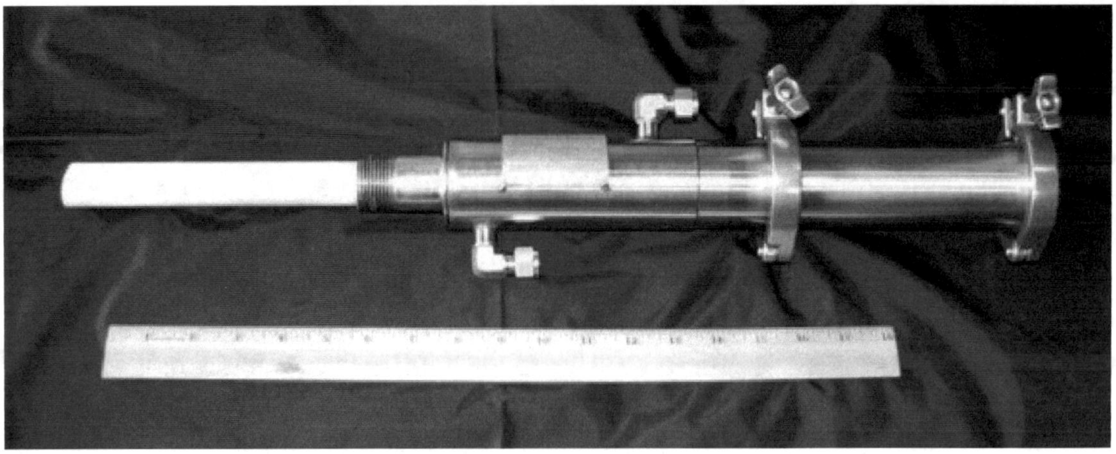

Figure 1. Prototype microwave antenna module for chemical looping.

The continuous wave microwave Doppler approach has been adopted for the initial prototype. Since the system has a short working distance, and is low power, this allows for a simple electronic configuration. A diagram of the system is shown in Figure 2. A stabilized frequency oscillator provides the microwave carrier signal at frequency f0 (10 GHz for the prototype). The primary signal is passed through a circulator to the antenna (shown as a horn antenna in the figure). The reflected signal is Doppler shifted by a small amount, Δf, and is received by the same antenna. The received signal passes back through the circulator to the downmixer. A portion of the original carrier signal is sampled through a directional coupler, attenuated, and sent to the downmixer, where it is mixed with the return signal to produce the baseband signal at frequency Δf. The baseband signal is lowpass filtered before analog-to-digital conversion, and processing on a computer.

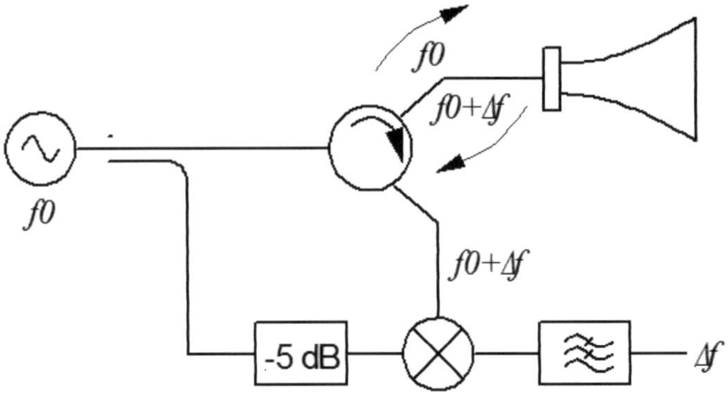

Figure 2. Diagram of the microwave electronics used for prototype.

The reflected wave from an object moving at velocity *v* interacting with a microwave beam at frequency *f* is shifted by

$$\Delta f = 2f\frac{v}{c}\cos\theta$$

where θ is the angle between the direction of propagation of the beam and the object velocity vector and c is the speed of light. For a microwave carrier frequency of 10 GHz and velocities of the order of a few meters/sec the frequency shift is a few hundred Hertz. The angle θ is selected near 45 degrees because a large frequency shift is desired, but small angles are geometrically impractical to implement in hardware.

In the solids flow sensing application, the reflection is from a large number of particles, resulting in a distribution of frequencies in the reflected signal. To quantify the distribution and relate it to the velocity, the following approach has been taken (see Figure 3). First, for a set of data, the fast Fourier transform is used to compute the frequency spectrum. The frequency components less than f_{min} are set to zero, to reduce the impact of 1/f noise. The frequency spectrum is truncated at f_{max} to exclude components that correspond to unphysically large particle velocities. Then the cumulative distribution function (CDF) is formed from the modified frequency spectrum. The maximum value D_{max} of the cumulative distribution function is a measure of the total reflected power. The frequency corresponding to a CDF value of half of D_{max} is found as a representative value of the average Doppler shift, and therefore particle velocity.

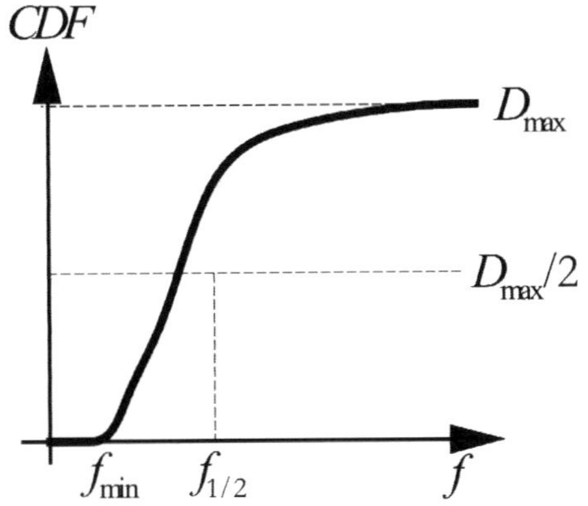

Figure 3. Parameters used to estimate flow rate from cumulative distribution function (CDF) of the Doppler shifted microwave signal.

For lab testing and calibration, the approach taken has been to use a rotating table feeder to provide a controlled particle flow past the sensor. Particulate material is fed through a drop tube to the surface of a small round disk (the table), near its edge. The particles dropped onto the rotating table rotate away from the drop tube, and are brushed off the table into a funnel. The rotation of the disk and the height of the gap between the drop tube and the disk control the particle feed rate. The particles drop from the funnel, gain velocity from gravitational acceleration, and pass by the sensor under test. The particles are collected at the bottom of the drop tube and continuously monitored with a load cell to determine the actual mass vs time (solid flow rate) for calibration. Within a limited range, the solids flow rate produced is steady.

Additional testing of the microwave Doppler sensor is being performed on other experimental apparatuses at NETL. The cold flow model of the Chemical Looping Reactor allows visual observation of gas-solids flow behavior, albeit at room temperature. The High Temperature Solids Flow Verification apparatus has been constructed to provide a high temperature batch test capability. It provides a short period of gravity driven flow of high temperature particles, which have been preheated to CLC temperature conditions. Lastly, testing on the Chemical Looping Reactor provides experience with a pilot scale application environment.

Experiments

The table feeder was used to initially test the microwave Doppler sensor at low temperature, with hematite. The hematite was collected in a bucket suspended by a load cell to provide the reference measurement of accumulated mass, at the same time as the microwave sensor measurement. An example of the data for two different mass flow conditions is shown in Figure 4, where the table rotation speed was 10 rpm for the low flow case, and 25 rpm for the high flow case. The flow rate appears to be steady in each case. The microwave Doppler sensor was installed on the drop tube below the table feeder, and above the collection bucket, and used the prototype antenna module. The microwave Doppler sensor parameter D_{max} is shown for the same cases in Figure 5. While D_{max} is noisy, it does show a clear change with the change in solids flow rate.

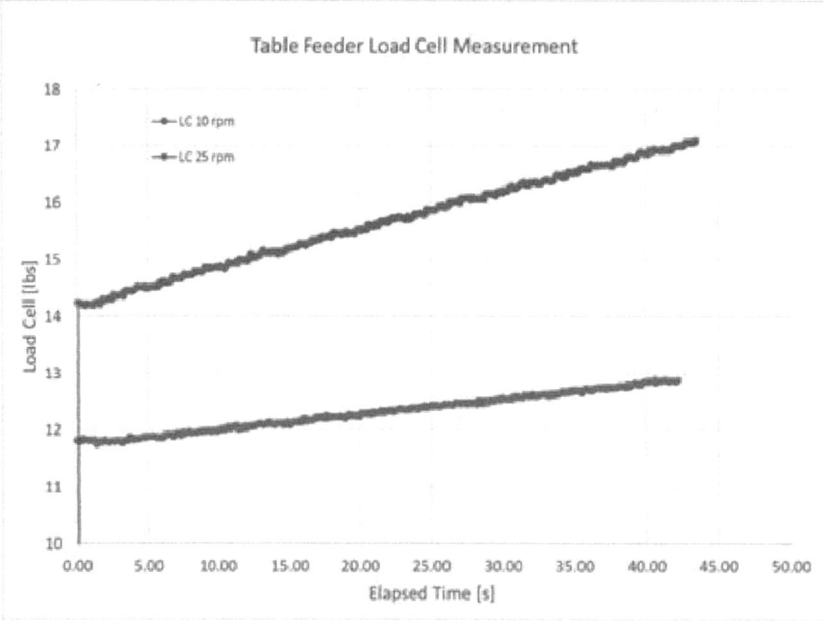

Figure 4. Accumulated mass measured by load cell during table feeder operation.

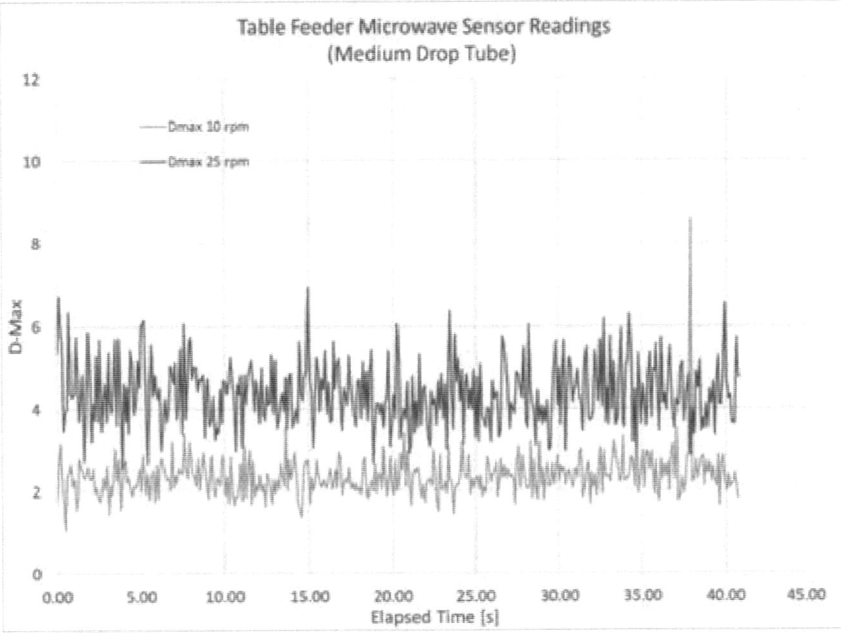

Figure 5. Microwave sensor parameter Dmax during table feeder operations..

The table feeder results show the validity of the general approach, and provided a preliminary room temperature calibration; however, the target application is at high temperature. To get early feedback on performance in application conditions, the prototype microwave Doppler sensor and antenna module have been installed during test operations of the Chemical Looping Reactor. While the antenna module has demonstrated mechanical resistance to the temperature and pressure conditions in the CLR, signal levels were low, and the data collected during the test runs has not been clearly correlated with the operating conditions indicated by the other system instrumentation. A laboratory bench test of a comparable length of the alumina rod material used as a high temperature waveguide in the antenna module found the microwave transmission of the alumina rod to decrease substantially as the temperature was increased above 800°C, as shown in Figure 6. The loss in the solids flow sensor application would be worse, due to the losses in both transmit and receive passes through the waveguide. In addition, during the most recent CLR operations, the end of the alumina waveguide was hard coated with hematite, which appears to have further reduced the microwave signal. It is postulated that this coating was due to a few higher temperature periods early in the test.

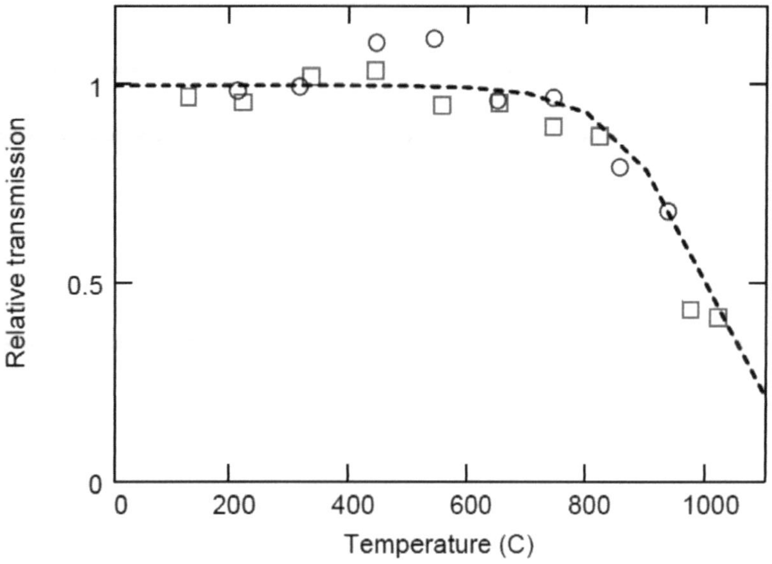

Figure 6. Relative transmission of an alumina rod at 10 GHz, heated by an electric tube furnace. Red symbols are measurements during heat up, and blue during cool down.

Several options are being explored to address the low signal problem observed during the CLR operations. First, the prototype microwave carrier signal power will be increased from 10 mW to at least 100 mW. Second, the signal-to-noise will be improved through additional amplification of the received signal. A third option is to change the design of the antenna module, including the waveguide material. Geometrical changes to the antenna module are also being explored [10].

Conclusions

NETL is working to address the technical challenges posed by the chemical looping combustion process. Part of the research effort is focused on sensor development, where measurement of solids flow in the high temperature process is a difficult challenge. A prototype sensor using a microwave Doppler radar approach has been constructed and tested at room temperature, and at high temperature in the Chemical Looping Reactor. While the prototype microwave antenna module showed sufficient durability for the pressures and temperatures of the application environment, the measurements in the CLR had poor signal-to-noise ratio, and did not correlate well with other instrumentation. Several possible modifications to the prototype have been identified to improve the signal-to-noise ratio.

Acknowledgements

This project was funded by the Department of Energy, National Energy Technology Laboratory, an agency of the United States Government. Mr. Charley was funded through a support contract with AECOM, Inc. Dr. Greve was supported in part by through the ORISE Faculty Research Program. Neither the United States Government nor any agency thereof, nor any of their employees, nor AECOM, Inc., nor any of their employees, makes any warranty, expressed or implied, or assumes any legal liability or responsibility for the accuracy, completeness, or usefulness of any information, apparatus, product, or process disclosed, or represents that its use would not infringe privately owned rights. Reference herein to any specific commercial product, process, or service by trade name, trademark, manufacturer, or otherwise, does not necessarily constitute or imply its endorsement, recommendation, or favoring by the United States Government or any agency thereof. The views and opinions of authors expressed herein do not necessarily state or reflect those of the United States Government or any agency thereof.

References

1. Ekström, C., et al. (2009), "Techno-Economic Evaluations and Benchmarking of Pre-combustion CO2 Capture and Oxy-fuel Processes Developed in the European ENCAP Project," Energy Procedia, Volume 1, Issue 1, pp. 4233-4240.
2. Alstom Power Inc., Power Plant Laboratories, "Greenhouse Gas Emissions Control by Oxygen Firing in Circulating Fluidized Bed Boilers: Phase 1 – A Preliminary Systems Evaluation," Vol 1., DE-FC26-01NT41146, http://www.netl.doe.gov/File%20Library/Research/Coal/ewr/co2/41146-Alstom_Power_Final-I_2003_oxycombustion-CFB_R01_Volume.pdf
3. Berguerand, N., and Lyngfelt, A. (2008), "Design and Operation of a 10 kWth Chemical-Looping Combustor for Solid Fuels - Testing with South African Coal," Fuel, v. 87, pp. 2713-2726.
4. Shen, L., Wu, J. and X. Jun (2009), "Experiments on Chemical Looping Combustion of Coal with a NiO Based Oxygen Carrier," Combustion and Flame, v. 156, 721-728.
5. Lozza, Giovanni, et al. (2006), "Three Reactors Chemical Looping Combustion for High Efficiency Electricity Generation with CO2 Capture from Natural Gas," in Proceedings of ASME Turbo Expo 2006: Power for Land, Sea and Air, Barcelona, Spain, GT2006-90345.
6. Consonni, S., et al. (2004), "Chemical-Looping Combustion for Combined Cycles with CO2 Capture," in Proceedings of ASME Turbo Expo 2004: Power for Land, Sea and Air, Vienna, Austria, GT2004-53503.
7. F. Gao, D. W. Greve, and I. J. Oppenheim (2012) "Launcher design for chemical looping combustion," in COMSOL Conference, Boston, Oct. 3-5, 2012.

8. F. Zhang, D. W. Greve, F. Gao, and I.J. Oppenheim (2012) "Doppler solid flow measurement for chemical looping combustion," in 2nd Global Conference on Microwave Energy Applications, Long Beach, California, July 23-27.
9. Greve, D. W., Oppenheim, I. J., Chorpening, B. T. and J. Charley (2013), "Microwave Doppler flow sensor for chemical looping combustion systems," IEEE Sensors 2013, Baltimore, Nov. 3-6.
10. Greve, D. W., Oppenheim, I. J., Chorpening, B. T. and J. Charley (2015), "Doppler sensing of particle velocity in mixed-phase flows," 2015 Workshop on Multiphase Flow Science, Morgantown, WV, Aug. 12-13.

Novel process for energy storage and conversion: combined chemical looping

Vladimir V. Galvita, Hilde Poelman and Guy B. Marin

Laboratory for Chemical Technology, Ghent University, Technologiepark 914, B-9052 Ghent, Belgium, E-mail: Vladimir.Galvita@Ugent.be

The development of reliable and environmentally friendly approaches for energy conversion and storage is one of the key challenges that our society is facing [1, 2]. Energy can be stored in different forms: in an electric or magnetic field; as mechanical energy; or as chemical energy [3]. In a typical energy storage process, one type of energy is converted into another form which can easily be stored and converted for use when needed [4].

Chemical looping combustion (CLC) is an emerging combustion technology. In this process, fuel is oxidized by a reducible metal oxide and the reduced metal oxide is re-oxidized by air in a separate step [5, 6]. CLC hence produces a pure CO_2 stream, not diluted by N_2. By replacing air with H_2O or CO_2 as oxidizer, the chemical looping analogue to steam (CLSR) and dry reforming (CLDR) is obtained [6-8]. The latter converts CO_2 to high-purity CO, providing an efficient path for CO_2 conversion. The overall stoichiometry of this process indeed demonstrates that the oxidation step converts more CO_2than was produced in the fuel combustion step.

The calcium looping technology is a promising new technique for high-temperature scouring of CO_2 from flue gas and syngas. Calcium looping cycles have been intensively investigated in order to produce a concentrated CO_2 stream from the utilization of fossil fuels and biomass. Efficient CO_2 capture and storage using Ca-based sorbents can be achieved via the reversible reaction $CaO + CO_2$ $CaCO_3$ ($\Delta H_{298\,K} = -178$kJ mol^{-1}), the so-called Ca-looping cycle [9].

The present novel concept proposes energy storage and conversion based on a combination of chemical and calcium looping processes [10]. It uses a physical mixture of a CO_2 sorbent and an oxygen storage material. CO_2 serves as mediation gas to facilitate metal oxidation and carbon gasification into CO by means of chemical looping, while the calcium looping process ensures storage and release of CO_2.

Combination of Fe_3O_4/Fe redox cycles with $CaO/CaCO_3$ looping in this novel concept of "combined chemical looping" provides a number of important advantages. Where chemical looping dry reforming immediately oxidizes fuel to CO_2, which is further converted to CO, the CO_2 mediation gas is now stored to be used at will. The $CaCO_3$ solid material acts both as CO_2 reservoir and supplier, by cyclic $CaO/CaCO_3$ calcination-carbonation. In addition, the adsorption of CO_2gas from the stream during iron oxide reduction promotes faster material reduction and carbon formation.

The working principle of the "combined chemical looping" can be described as follows [10]. During the reduction step ("charge process"), $CH_4 + CO_2$ is fed into a mechanical mixture of Fe_3O_4 and CaO. Interaction of methane with Fe_3O_4 leads to formation of metallic iron and surface carbon as well as CO_2, H_2O and H_2 (Eq. 1 and 2). The H_2 (Eq. 2) obtained can be used for fuel cells. At the same time carbonation of calcium oxide occurs by interaction of CO_2 with CaO (Eq. 3). CaO particles react with CO_2 at typical high temperatures (600–700°C) to form $CaCO_3$.

$$CH_4 + Fe_3O_4 \rightarrow 2H_2O + CO_2 + 3Fe \quad (1)$$

$$CH_4 \rightarrow C + 2H_2 \quad (2)$$

$$CaO + CO_2 \rightarrow CaCO_3 \quad (3)$$

The material can be kept in this "charged" condition if storage is required or be put to immediate use. For the oxidation step ("discharge"), the temperature of the sample is increased by 50-150°C which leads to decomposition of calcium carbonate into CaO and CO_2 (Eq. 4).

$$CaCO_3 \rightarrow CaO + CO_2 \quad (4)$$

At the same time, the interaction of $CO_2$with metallic iron as well as with carbon produces CO via the following chemical reactions:

$$4CO_2 + 3Fe \rightarrow Fe_3O_4 + 4CO \quad (5)$$

$$C + CO_2 \rightarrow 2CO \quad (6)$$

The thus generated CO could for example be used in a solid oxide fuel cell (SOFC) where it is electrochemically oxidized,

$CO + O^{2-} = CO_2 + 2e^-$, producing electricity and CO_2.

The objective of the present study is the experimental validation of the proposed concept of combined chemical looping. The proof of concept was performed in two steps. In the first, a single NiO-Fe_2O_3/CeO_2 material was used and CH_4 was fed for iron oxide reduction and carbon formation. Interaction of this material with CO_2 led to CO formation by iron and carbon oxidation by CO_2. In the second step, the NiO-Fe_2O_3/CeO_2 material was mixed with CaO to perform a complete combined chemical looping test with a feed of CH_4+CO_2. The amount of CO produced is higher than the theoretical amount for Fe_3O_4, because carbon deposits from CH_4equally contribute to the CO yield. It was shown that this cycle of "charge-discharge" can be repeated with limited decrease of the rate of the involved reduction/oxidation steps. The proposed methodology offers a wide flexibility both in energy storage/release and fuel type. For charging, diluted gas streams can be applied, making it very attractive for present-day applications.

Acknowledgments

This work was supported by the "Long Term Structural Methusalem Funding by the Flemish Government", the Fund for Scientific Research Flanders (FWO, project number 3G004613), and the Interuniversity Attraction Poles Programme, IAP7/5, Belgian State – Belgian Science Policy.

[1] A.M. Omer, Renew. Sust. Energ. Rev. 12 (2008) 2265-2300.

[2] A. Antonino Salvatore, P. Bruce, B. Scrosati, J.-M. Tarascon, W. van Schalkwijk, Nat Mater 4 (2005) 366-377.

[3] L. Schlapbach, A. Zuttel, Nature 414 (2001) 353-358.

[4] T. Kousksou, P. Bruel, A. Jamil, T. El Rhafiki, Y. Zeraouli, Energy storage: Applications and challenges, 2014, pp. 59-80.

[5] S. Bhavsar, M. Najera, R. Solunke, G. Veser, Catalysis Today 228 (2014) 96-105.

[6] L.-S. Fan, L. Zeng, S. Luo, AIChE Journal 61 (2015) 2-22.

[7] V.V. Galvita, H. Poelman, C. Detavernier, G.B. Marin, Applied Catalysis B: Environmental 164 (2015) 184-191.

[8] V.V. Galvita, H. Poelman, V. Bliznuk, C. Detavernier, G.B. Marin, Industrial & Engineering Chemistry Research 52 (2013) 8416-8426.

[9] E.J. Anthony, Greenhouse Gases: Science and Technology 1 (2011) 36-47.

[10] V.V. Galvita, H. Poelman, G.B. Marin, Journal of Power Sources 286 (2015) 362-370.

434584 The Syngas Chemical Looping and Coal Direct Chemical Looping Processes for Hydrogen and Electricity Production and Carbon Capture: Process Development and Pilot-Scale Demonstration

Abstract:

Chemical looping technologies have evolved into a promising alternative for the efficient conversion of carbonaceous fuels to electricity and/or high value chemicals with minimal carbon emissions. These processes utilize an oxygen carrier solid material to indirectly supply oxygen to the fuel source eliminating the need for costly and energy intensive gas-gas separation techniques. In the case of power production, the chemical looping scheme allows for high exergy efficiency, as the high grade heat produced from the oxidation of the oxygen carrier can be extracted for steam production, while the lower grade heat is recuperated in the oxygen carrier to perform its endothermic reduction. With the exponential growth of research and publications in this field, chemical looping has expanded to encompass power and chemical production with in-situ gas separation. The Ohio State University (OSU) has successfully developed the syngas chemical looping (SCL) and coal direct chemical looping (CDCL) processes from laboratory studies to two fully integrated pilot scale demonstration facilities for the conversion of gaseous and solid fuels, respectively, to high purity H_2 and heat with nearly 100% CO_2 capture.

The present paper summarizes key developments in the SCL and CDCL processes Specifically, TGA tests on the oxygen carrier recyclability and reactivity as well as jet cup and crush strength tests on the particle attrition resistance are provided. The results from over 1,000 hours of operation of two 25 kW_{th} sub-pilot units showing nearly 100% conversion of 7 gaseous and solid fuel types to CO_2 are summarized. A comprehensive techno-economic analysis of the CDCL was completed with Babcock & Wilcox Power Generation Group. The results indicate the CDCL process for power generation can exceed Department of Energy's target of <35% increase in cost of electricity with greater than 90% carbon capture. An overview of the SCL pilot unit design, construction, and commissioning is provided. The operation results of the pilot unit under high pressure and temperature conditions are also discussed.

435495 Analysis of Oxygen Carrier Reactivity and Fuel Conversion in Copper-Based Chemical Looping with Oxygen Uncoupling (CLOU) Systems

Abstract:

2015 AIChE Annual Meeting- "Chemical Looping Processes"

Performance of a 10 kW_{th} Dual Fluidized Bed Chemical Looping Combustion Reactor Operating in iG-CLC and CLOU Modes

Kirsten M. Merrett and Kevin J. Whitty

Dept. of Chemical Engineering, University of Utah, Salt Lake City, UT

ABSTRACT

Chemical looping combustion (CLC) is a technology to combust solid fuels while intrinsically separating the greenhouse gas CO_2. This eliminates the energy cost associated with gas separation, making CLC an attractive technology for CO_2 separation. CLC involves circulating a metal-based "oxygen carrier" between two reactors in which it is oxidized by air and then reduced by a fuel, in essence combusting the fuel. In the variant of CLC known as chemical looping with oxygen uncoupling (CLOU), the oxygen carrier releases gas-phase oxygen as O_2 once in the fuel reactors reducing environment, allowing for a gas-solid reaction to occur. Particles are then circulated to the air reactor and reoxidized. The natural mineral ilmenite presents a cost-effective oxygen carrier material with demonstrated strength and cyclability, while copper allows for the desired oxygen uncoupling behavior of the metal oxide during CLOU operation.

The University of Utah has developed a 10 kW_{th} bench scale dual fluidized bed chemical looping combustion system. The goal of the study presented here was to assess system performance and compare conventional chemical looping combustion, known as indirect gasification-CLC or iG-CLC, with CLOU chemical looping. In this study two different fuels, coal and coal char, were processed and the degree of fuel conversion, CO_2 capture efficiency and CO_2 purity were evaluated. Performance of the unit itself is also evaluated by changing variables such as temperature, flow rate, bed inventory, and circulation rate to determine optimal performance.

Driving Innovation ♦ Delivering Results

NETL's 50kW Circulating Chemical Looping Combustion Test Facility and Operating Experiences

Doug Straub, Justin Weber, and Dr. Jim Spenik

November 11, 2015

U.S. DEPARTMENT OF ENERGY | National Energy Technology Laboratory

Presentation Outline

- **Motivation/Background**
  - Why is there interest in chemical looping combustion?
- **Test facility description**
  - Background
  - Operating modes
- **Preliminary test results**
- **Conclusions and future work**

DOE/NETL Sites

Motivation for this work

- **Presidential GHG reduction goals**
- **Domestic importance of fossil fuels**
  - Need fossil fuel options to reduce GHGs
- **CLC technology has "potential" to achieve DOE goals**

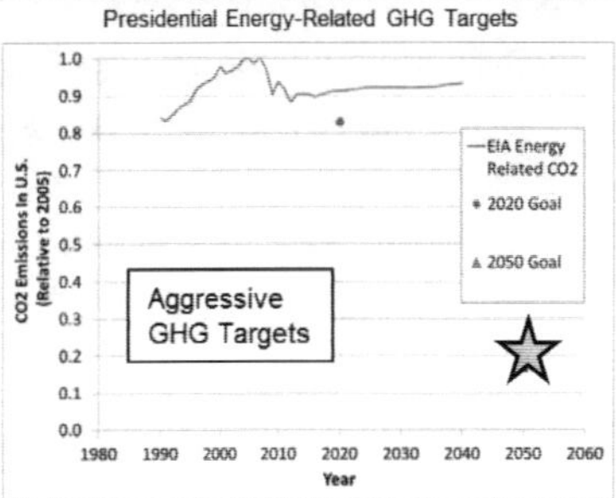

http://www.eia.gov/environment/emissions/carbon

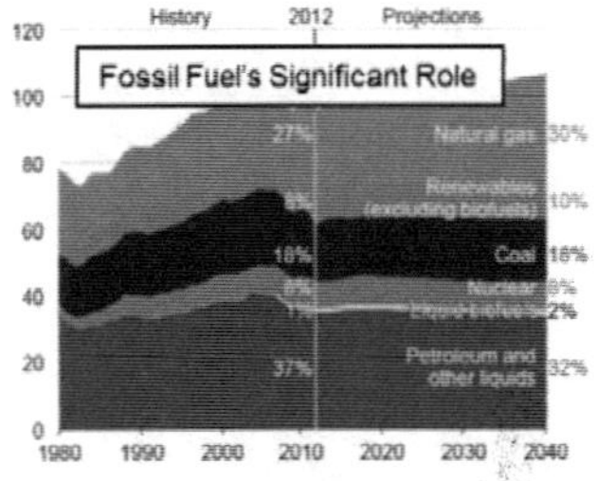

AEO2014 – Early Release

Exhibit ES-3 Cost of electricity breakdown comparison

Cost	Fe_2O_3 (\$/MWh)	$CaSO_4$ (\$/MWh)	Conventional PC BBR Case 12
Capital	49.6	53.4	73.1
Fixed	11.3	12.2	15.7
Variable	25.7	8.4	13.2
Maintenance materials	3.2	3.5	4.7
Water	0.4	0.4	0.9
Oxygen carrier makeup *	18.7	1.1	N/A
Other chemicals & catalyst	1.9	1.7	6.4
Waste disposal	1.4	1.7	1.3
Fuel	28.4	30.8	35.3
Total	115.1	104.7	137.3

*Fe_2O_3 oxygen carrier makeup: 132 tons/day @ \$2,000 per ton; Limestone carrier makeup: 439 tons/day @ \$33.5 per ton

Ref: U.S. Department of Energy (DOE), National Energy Technology Laboratory (NETL). Guidance for NETL's Oxycombustion R&D Program: Chemical Looping Combustion Reference Plant Designs and Sensitivity Studies. Pittsburgh : s.n., 2014. DOE/NETL-2014/1643

NETL/ORD's Chemical Looping Combustion

What is our end goal?

- **Determine if CLC is a feasible technology for FE and worthy of additional investment/development**
  → Data and information for strategic decision making
- **If it is feasible, THEN**
  - Help developers overcome technical issues
  - Help technology be successful
  - Ultimately commercialization
  → jobs and growth

NETL/ORD – Key Components of CLC Effort

CFD tools: Technology scaling

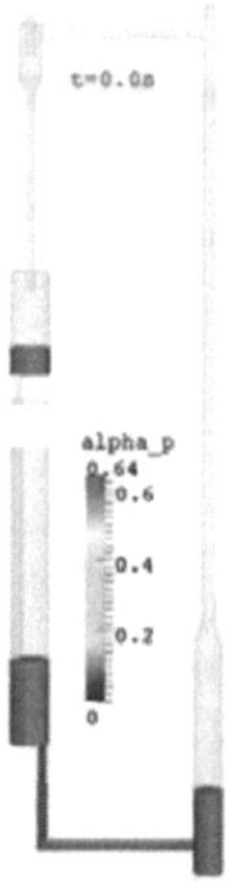

Sensors: Enable reliable systems

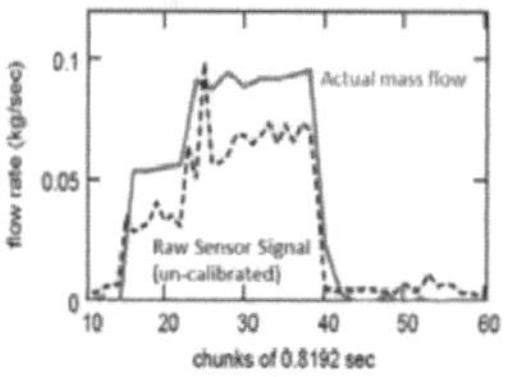

Carrier materials: Durability, cost, performance

NETL-CLC facilities: Expertise and benchmarking

Nomenclature For NETL's 50kW Test Facility

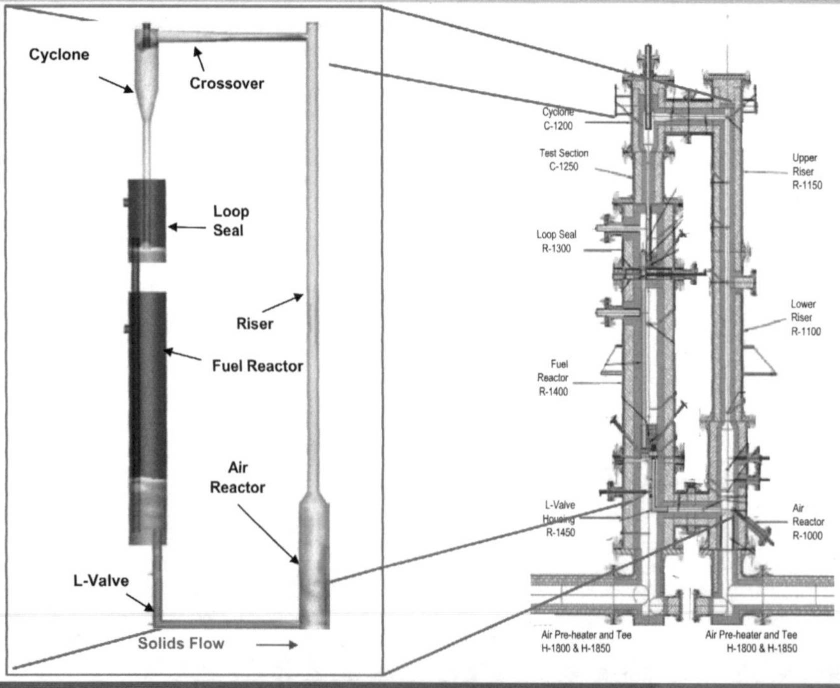

Chemical Looping Flow Diagram

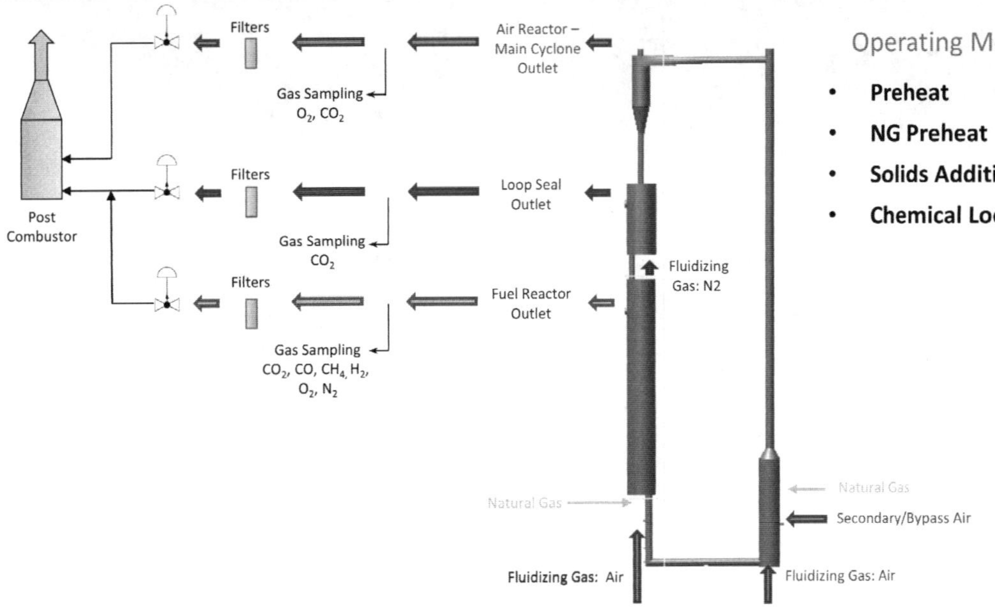

Operating Modes

- **Preheat**
- **NG Preheat**
- **Solids Addition**
- **Chemical Looping**

Operating Modes

- **Electric preheat**
  - Room temperature → Auto-ignition temperature
- **Natural gas augmented preheat**
  - 1200F to 2000F
  - Gas phase combustion in both reactors
- **Carrier addition**
  - Reduce gas flows
  - Add carrier in batches via lockhopper
- **Chemical looping combustion**
  - Transition from air to N2 as fluidizing gas in FR
  - Adjust natural gas flow for CLC

Oxygen Carrier Materials

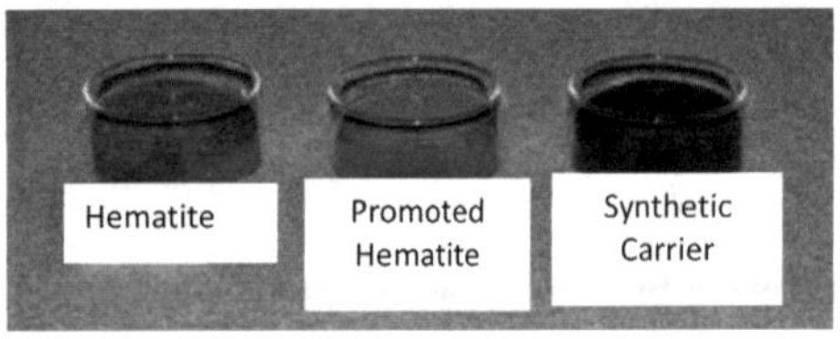

Pre-Test

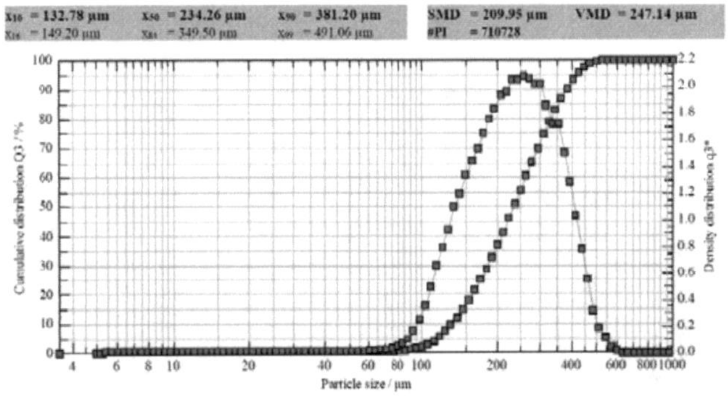

Post-Test

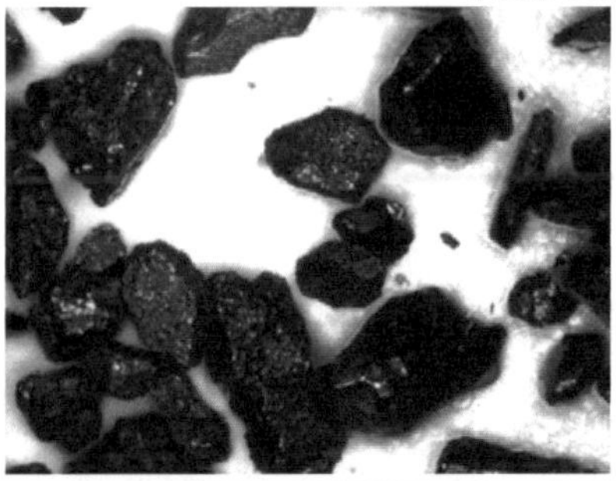

Process Data from NETL Unit – Promoted Hematite Carrier – 1st Test

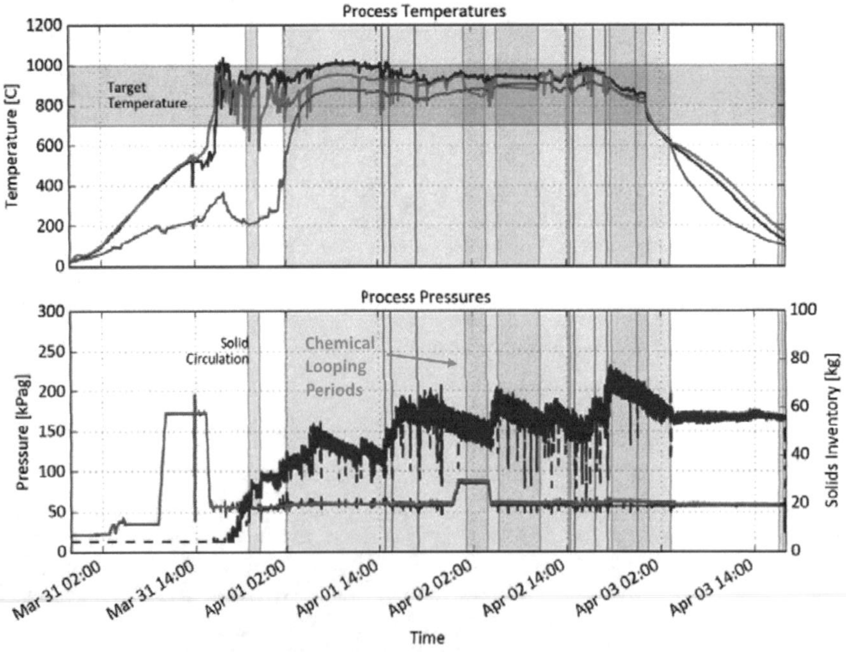

Promoted Hematite – 1st Test

- **Over 12 hours of CL operation**
- **No significant process upsets**
- **Carbon balance was less than 90%**
  - Internal leakage through refractory in L-valve region
- **Fuel conversion typically less than 50%**

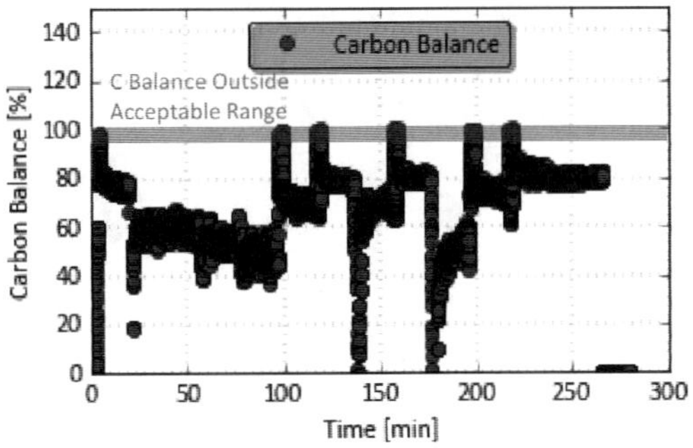

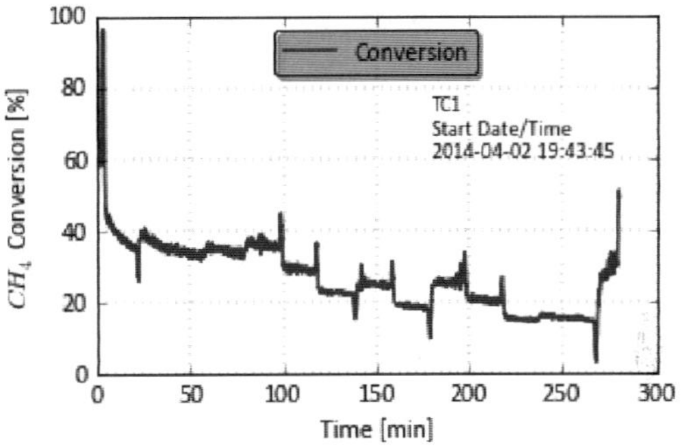

Process Data from NETL Unit – Promoted Hematite Carrier – 2nd Test

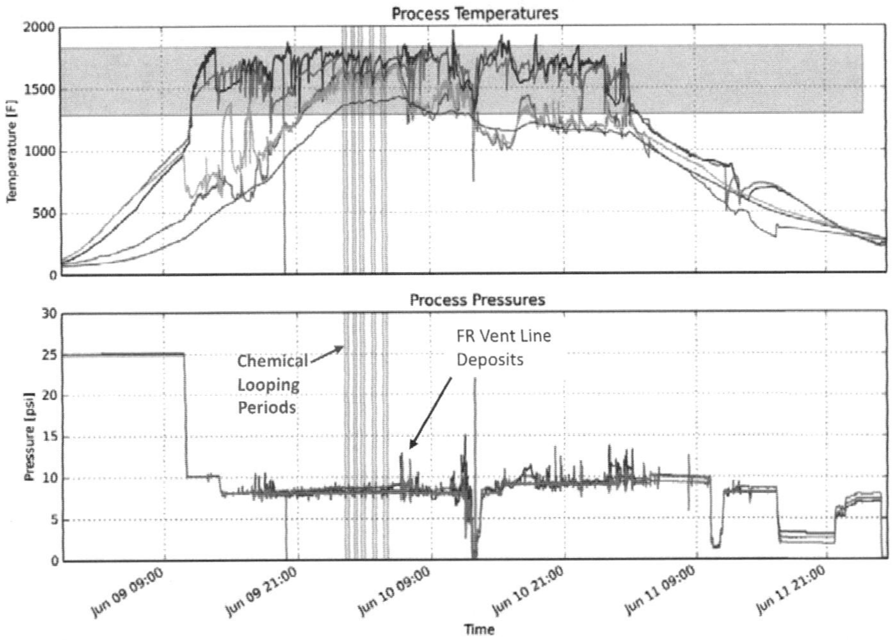

Promoted Hematite – 2nd Test

- **Carbon balance was significantly improved**
  - Greater than 90%
- **Fuel conversion still typically less than 50%**
- **Short duration test periods**
- **Operational issues encountered**

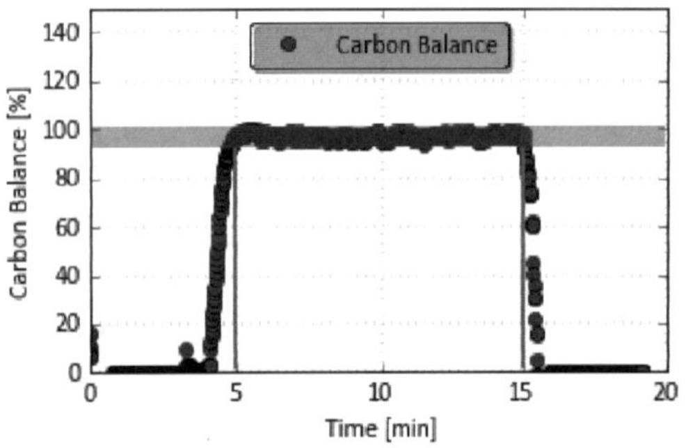

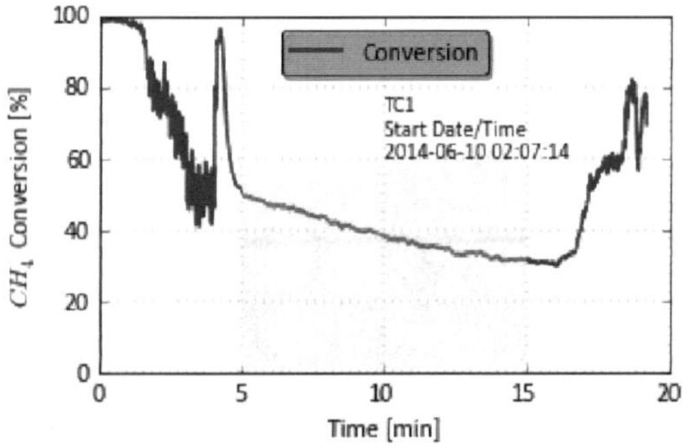

Promoted Hematite – 2nd Test

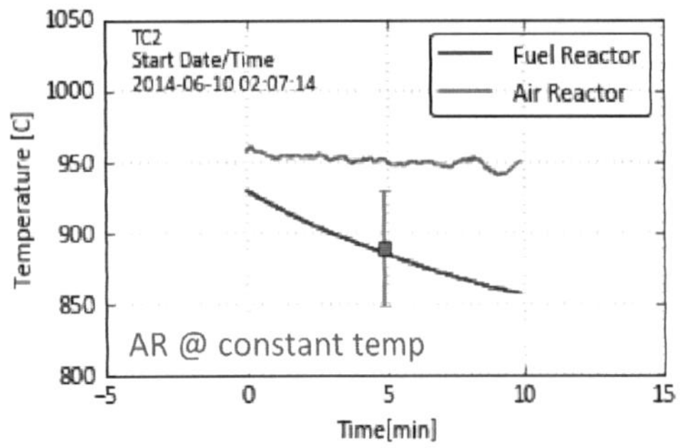

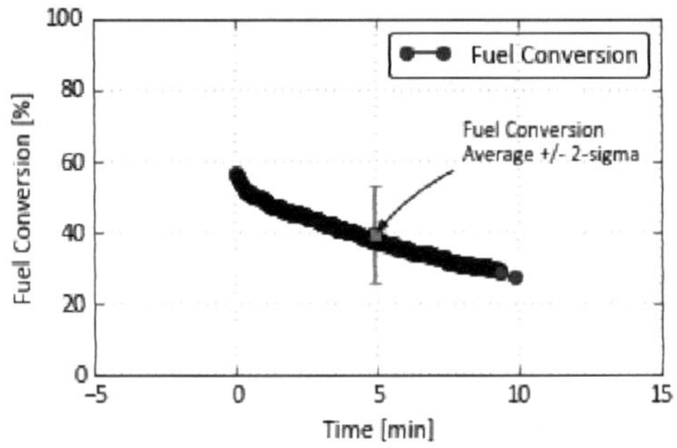

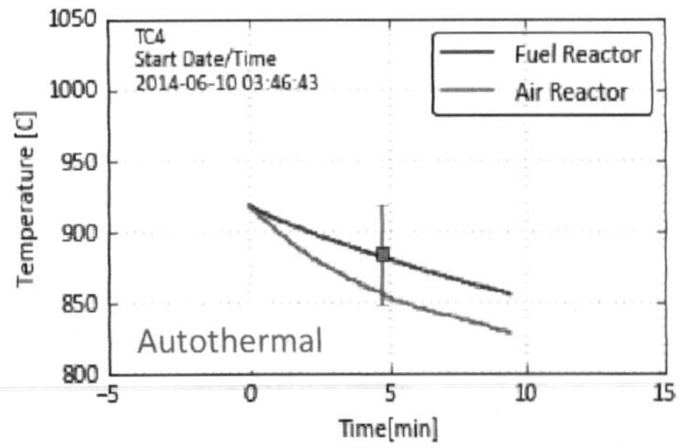

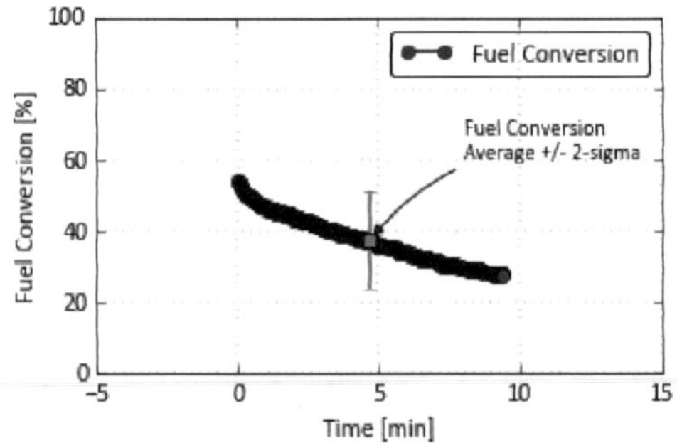

Source of Operational Issues

- **Cake-like deposits have been seen in tests with MgO promoted hematite carrier**
  - Deposits only in the FR vent lines (reducing environment)
  - Downstream/lower temperature components
- **Produced enough backpressure to affect pressure balance and control**

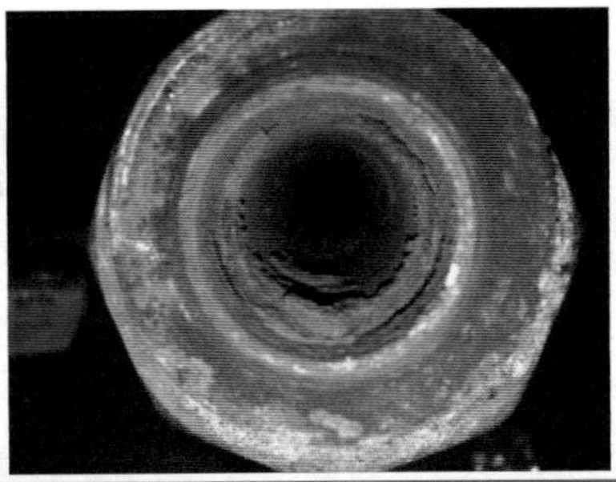

U.S. DEPARTMENT OF ENERGY | National Energy Technology Laboratory 15

Data from Hematite – 1st Test

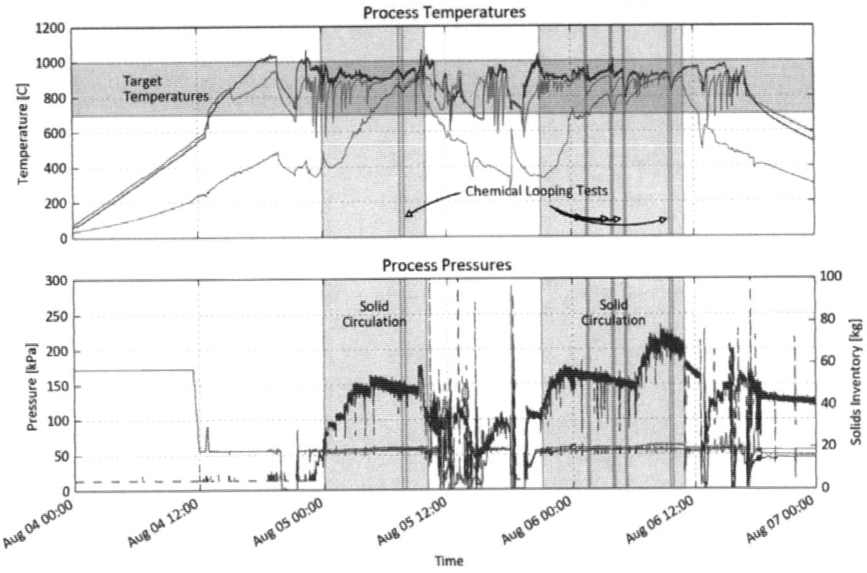

Operational issues and process upsets were encountered

U.S. DEPARTMENT OF ENERGY | National Energy Technology Laboratory 16

Process Upset → Sudden Loss of Solids

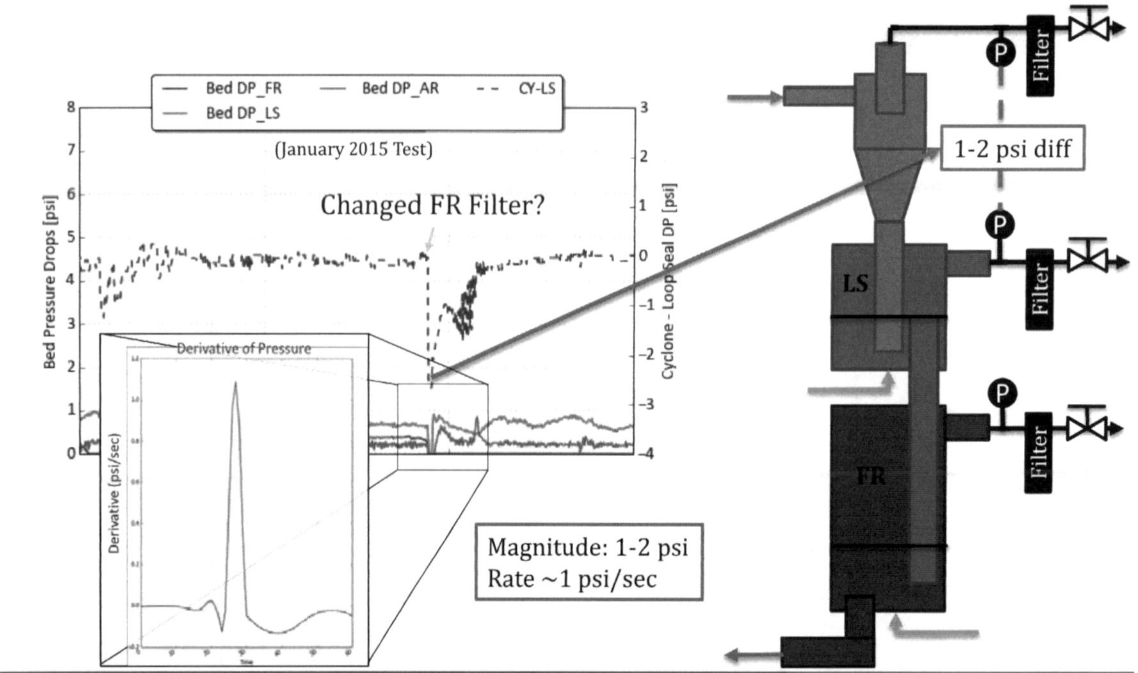

Summary

- **CLC technology has "potential" to achieve cost-effective reduction in GHG emissions from fossil-fueled combustion**
- **NETL's 50kW$_{th}$ CLC test facility**
  - Bubbling fluidized bed fuel reactor
  - Bubbling/turbulent fluidized bed air reactor
- **Tested several different oxygen carriers**
  - Hematite (Wabush Mine; Wabush, Newfoundland and Labrador, Canada)
  - MgO promoted hematite
- **Findings**
  - Mg promoted carrier materials produced coatings/cake in downstream lower temperature piping components
  - Operational pressure upsets have resulted in high rates of oxygen carrier loss
    - Pressure perturbations of 1-2 psi at rates of 1 psi/second have been recorded

Future Efforts

- **Establish baseline carrier loss/make-up requirements for this test rig**
  - Sensitivity studies have shown this to be a critical performance parameter
  - Need better solids mass closure
- **High temperature microwave doppler solids circulation sensor**
- **Collaborations on oxygen carrier testing are encouraged**
  - NETL attrition test unit would be used to screen materials
  - Quantity should be in the 100-150 kg range for one test

U.S. DEPARTMENT OF ENERGY | National Energy Technology Laboratory 19

In memory of James Spenik, Co-author, Colleague and Friend

It's All About a Clean, Affordable Energy Future
For More Information, Contact
Doug Straub (304)285-5444;
Douglas.straub@netl.doe.gov
Delivering Yesterday and Preparing for Tomorrow
U.S. DEPARTMENT OF ENERGY
National Energy Technology Laboratory
21

428877 The Shale Gas-to-Syngas Chemical Looping Process for High Purity Syngas Production

Abstract:

The shale gas-to-syngas (STS) chemical looping processes is being developed at The Ohio State University for the production of high-purity syngas from shale gas. In contrast to conventional processes for syngas production, the STS process eliminates the need for additional steam or molecular oxygen from an air separation unit (ASU). The STS process utilizes a unique co-current moving bed gas-solid reducer reactor for the partial oxidation of gaseous fuels to syngas with metal oxide oxygen carriers. The reduced oxygen carriers from the reducer are regenerated in an oxidation operation with air in the combustor reactor. A modified Ellingham Diagram is developed to search for suitable metal oxides for the production of high-purity syngas. Various metal oxide compounds with desirable thermodynamic properties were identified. Through the combination of the suitable metal oxides and co-current moving bed reducer, high fuel conversion and syngas purity can be achieved. In this study, active iron-titanium composite metal oxide (ITCMO) materials are used as the oxygen carrier for the demonstration of the STS process. The co-current moving bed provides a desirable gas-solid contacting pattern that minimizes carbon deposition while maximizing the syngas yield. The syngas produced in the STS process can achieve a flexible H_2:CO molar ratio with little CO_2, unconverted CH_4, and steam, which is desirable for downstream processes to produce methanol, dimethyl ether, gasoline, and other fuels and chemicals. The experimental results for reaction kinetics including oxygen carrier recyclability and pressure effects were obtained from thermogravimetric analysis (TGA), and syngas generation results from fixed bed, bench-scale moving bed, and sub-pilot scale moving bed reactor demonstrations were achieved in this study. The bench and sub-pilot demonstrations confirmed the syngas produced from the STS process is consistent with the thermodynamic equilibrium for the reduced ITCMO material.

(580e) The Influence of Reactivation By Steam Hydration on Attrition of Sorbent Particles in Ca Looping Applications

Coppola, A.,
Montagnaro, F.,
SALATINO, P., Università degli Studi di Napoli Federico II
Scala, F., Università degli Studi di Napoli Federico II
Originally presented on: 11/11/2015 16:55:00 - 17:20:00

The present study addresses steam hydration as a reactivation technique of the CO_2 capture ability of spent limestone-based sorbents from fluidized bed calcium looping systems. A reference, high-calcium, reactive limestone was first deactivated by carrying out lab-scale fluidized bed calcium looping tests (calcination at 940°C in a 70% CO_2 atmosphere; carbonation at 650°C in a 15% CO_2 atmosphere), and then it was steam hydrated (at 250°C in a 50% steam atmosphere) in the same fluidized bed, for times ranging from 10 to 60 min. On-line flue gas analysis, continuous capture of the elutriated fines and analysis of particle size distribution were performed during additional calcium looping tests after sorbent reactivation. Thermogravimetric analysis, scanning electron microscopy and porosimetry were directed to characterize the microstructural features of the spent and the steam hydrated sorbents. Moreover, the different materials were subjected to ex situ impact fragmentation tests. In this way, it was possible to investigate the effect of the hydration time on the following aspects: the changes in the physico–chemical and microstructural properties induced by the hydration treatment; the reactivation of the sorbent CO_2 capture capacity; the attrition/fragmentation tendency of the reactivated materials.

It was observed that steam hydration followed by dehydration of the reactivated sorbent in the hot fluidized bed developed an increased porosity, hence improved the rate and the extent of the CO_2 uptake. At the same time also the attrition and fragmentation propensity of the reactivated sorbent was increased. For the given limestone, the optimal trade off between sorbent reactivity/uptake and mechanical strength was achieved after 60 min hydration of the spent sorbent, but it is expected that this result cannot be generalized, as it is critically dependent on the specific sorbent texture. The comparison between water and steam hydration of the spent sorbent as a reactivation means indicated that steam hydration is more favourable than water hydration. Albeit the latter gave rise to a larger water uptake, the prolonged soaking in liquid water made the reactivated sorbent more susceptible to attrition.

RAMAN GAS ANALYSIS FOR CHEMICAL LOOPING

Benjamin T. Chorpening, U. S. Department of Energy, National Energy Technology Laboratory, Morgantown, WV, USA
Michael P. Buric, U. S. Department of Energy, National Energy Technology Laboratory, Morgantown, WV
Steven D. Woodruff, U. S. Department of Energy, National Energy Technology Laboratory, Morgantown, WV
Justin M. Weber, U. S. Department of Energy, National Energy Technology Laboratory, Morgantown, WV
Douglas L. Straub, U. S. Department of Energy, National Energy Technology Laboratory, Morgantown, WV

Abstract

Chemical looping combustion is being developed by the U.S. Department of Energy as a possible technology for heat or power generation with integrated carbon dioxide separation. Improved sensors and controls are among the technical challenges that must be overcome for the commercial success of chemical looping. For improved control of the chemical looping process, accurate and rapid measurement of key process parameters is required. The Raman Gas Analyzer is being developed for real-time multi-gas analysis, to enable improved fossil energy process control. Application of the Raman Gas Analyzer to the Chemical Looping Reactor at NETL is reported.

Introduction

Chemical looping combustion is being developed with support of the U.S. Department of Energy because it has potential for carbon capture with less efficiency penalty than conventional power cycles with post-combustion capture [1,2]. Chemical looping uses an intermediary, such as a metal oxide, to collect oxygen from air in the air reactor. The oxidized particles are then transported to the fuel reactor, where the oxygen carrier particles are reduced. If complete combustion can be achieved, the product gas stream contains mostly carbon dioxide and water vapor, which can be readily separated for carbon capture, utilization and storage. An inert buffering gas can be injected in the process between the reactors to help separate the oxidation and reduction processes.

One of the challenges for operation of a chemical looping combustion process is the measurement of the gas compositions, which are indicative of the operating performance of those reactors. Based on these real-time measurements, solids inventory and gas flows may be adjusted to optimize the desired reactor performance. For process development, the produced gases also provide valuable information to characterize the performance of the oxygen carriers. While conventional gas analyzers can provide some of the data desired, the Raman Gas Analyzer is being developed to provide faster measurement of multiple gases simultaneously. At NETL, the Raman Gas Analyzer has been applied in the research operations of NETL's Chemical Looping Reactor test facility. The chemical looping reactor operates at positive pressure at high temperature (~ 1000 °C).

NETL's Chemical Looping Reactor (CLR) test facility is a small-scale circulating chemical looping combustion facility located at NETL's Morgantown, WV site. The NETL Chemical Looping Reactor has been constructed at the to characterize the performance of chemical looping oxygen carriers at a 50 kW_{th} scale, provide high fidelity process data to facilitate the development of computational tools, and provide a test bed for advanced instrumentation. It has been designed with a bubbling fluidized bed fuel reactor, and bubbling fluidized bed air reactor that transitions into a vertical riser to circulate the oxygen carrier material. The entire unit is refractory lined such that no external heat is added to the circulating solids flow path. The unit is currently configured to operate on natural gas fuel. The current operating temperature range can be varied between 700 – 1000°C and the typical operating pressure is approximately 150 kPa (1.5 bar). The performance of the fuel reactor is critical to the overall performance of the process. High fidelity gas composition monitoring of the fuel reactor effluent is critical to the CLR research goals.

Raman Gas Analyzer

The Raman Gas Analyzer (RGA) has been recently developed to be a fast, non-destructive instrument for on-line measurement of gas composition [3]. The Raman Gas Analyzer is capable of reporting the concentrations of multiple species simultaneously, with sampling times below one second. For process control applications in energy or chemical production, these response times are much more amenable to optimal control than other alternatives such as gas chromatographs. In the Raman Gas Analyzer, a hollow-core capillary waveguide with a reflective lining is used as a flow-through sample cell [4]. The effect of using such a waveguide in a Raman process is to integrate Raman photons along the length of the sample-filled waveguide, greatly improving the optical collection efficiency in gas applications. A pre-commercial field prototype has been constructed by NETL, and is being tested for potential use in energy applications such as coal gasification, turbine control, well-head monitoring for exploration or production, and non-conventional gas utilization, and chemical looping. The prototype used for the testing of the chemical looping reactor is shown in Figure 1.

Figure 1. Raman Gas Analyzer prototype.

Raman spectroscopy has long been limited to a laboratory-only technique because of the small magnitude of Raman scattering signals that can be collected from a gas with free-space optical configurations. While Raman microscopes for solid or liquid samples are common, Raman sensing for gases is much more difficult due to the low densities and low Raman cross-sections of most gases of interest. The early research on this project resulted in the implementation of a micro-bore sized capillary waveguide in an optimal configuration for collection of spontaneous Raman scattering inside the hollow waveguide core [4]. This configuration was shown to increase the collection efficiency of the scattered Raman light by a factor of 1000 or more versus free-space collection with conventional optics. Using this in an industrial instrument permits the rapid collection of Raman signals from gases with short optical integration times, and thus real-time or near real-time measurement rates. This advancement in the instrument configuration enables greater benefit from the positive features of Raman. Firstly, all gas species except the noble gases exhibit Raman scattering signals. Each of these signals is intrinsically identifiable since the vibration or rotation of molecular bonds gives rise to specific Raman emissions of discrete energy. While solids and liquids exhibit continuous and broad energy transitions, those resulting from the excitation of gases are narrow and easily identified. Furthermore, the diatomic molecules all exhibit well defined and well separated Raman lines which can be used for their characterization. Lastly, systems operating in the visible region produce larger signals (Raman signal power increases as wavelength decreases via a 4th power relationship), and are well suited for recording with common CCD detectors or cameras. These features make spontaneous Raman an attractive method for gas analysis [3]. Combination with the capillary waveguide to increase the signal allows the approach to be realized for fast gas analysis.

Chemical Looping Reactor Operations

The NETL Chemical Looping Reactor (CLR) facility has been designed and constructed at NETL for the purpose of research and development the chemical looping [5-7]. The composition of gases produced by the fuel reactor is a key performance indicator for understanding the operating condition, and enabling control decisions to be made for the system. The CLR is presently fueled with natural gas, but nitrogen is used to fluidize the oxygen carrier particles. In addition, during certain operations such as warmup, air may be supplied to the CLR for a preheat combustion mode with natural gas. As a result, the product gases from the fuel reactor include nitrogen, oxygen, carbon dioxide, carbon monoxide, water vapor, and unreacted methane. Hydrogen may also be produced under limited conditions. The Raman Gas Analyzer is well suited for monitoring this mixture of gases in a single instrument.

Data from a 12 hour period of the CLR test operations is shown in Figure 2, which plots carbon dioxide, methane, oxygen, and nitrogen concentration measurements from the fuel reactor. The RGA reports a gas composition every second during this period. The on-line gas chromatograph reports data every 2 minutes. Both show the response of the fuel reactor product gas composition to operational changes. Water vapor was detected by the RGA, but is not shown in the figure, because the sample gas clean-up system included water removal with a 5°C (40°F) condenser. As a result, the measured water vapor was primarily a function of water vapor saturation at the condenser temperature, instead of the CLR process. The sample pressure after cleanup was about 5 psig. This limited the signal strength of the Raman scattering; the RGA signal improves with increased sample pressure.

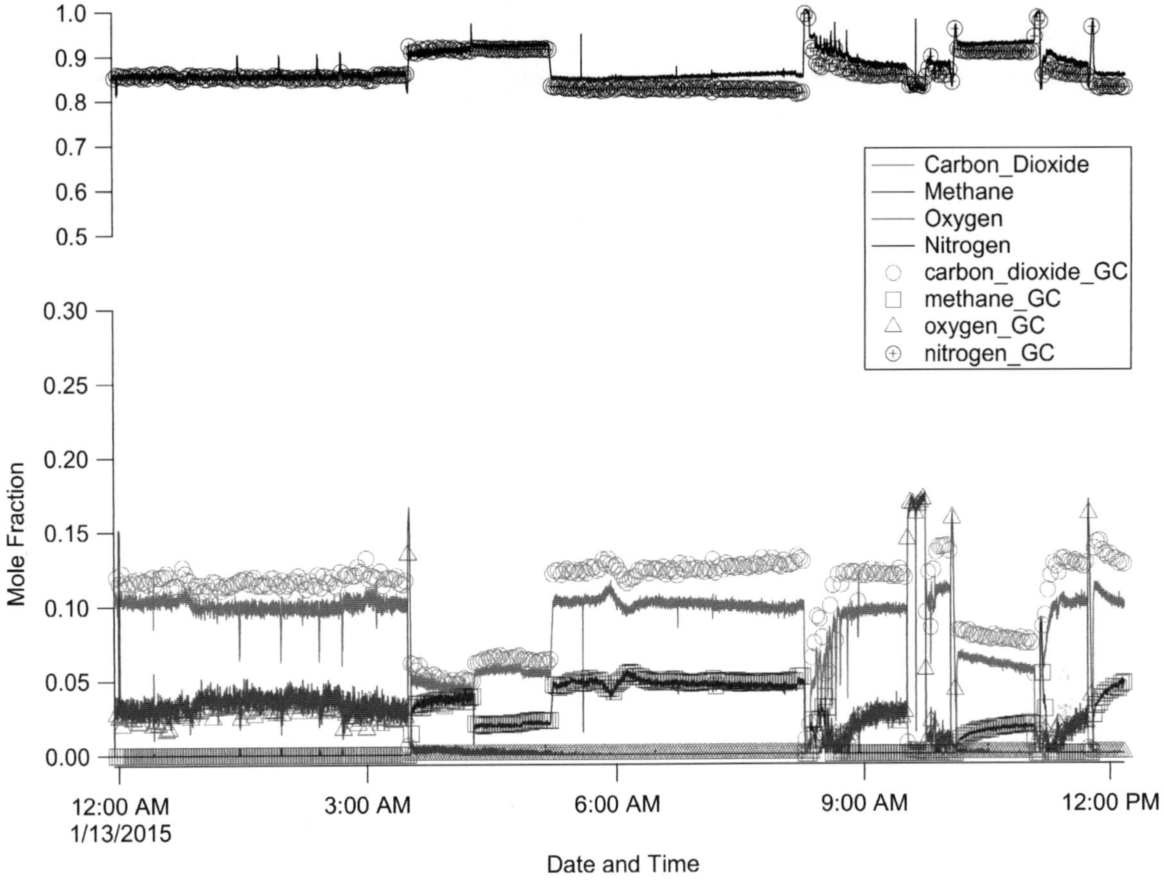

Figure 2. Raman Gas Analyzer (lines) and online gas chromatograph (open symbols) data from the fuel reactor during CLR research operations at NETL.

Although the two measurements are not exactly the same, Figure 2 does show that both RGA reported compositions and GC reported measurements generally agree on the operating behavior of the CLR fuel reactor. The most noticeable difference is in carbon dioxide concentration. Subsequent to the test run, it was found that the CO2 calibration gas mixture used for calibration of the GC contained substantially less CO2 than it was labeled (about 32% by volume, when labeled as 40%). This is thought to be the primary cause of the disagreement between the two instruments. Other potential causes include drift of the gas chromatograph, and a low signal-to-noise ratio on the RGA due to the low sample line pressure. During transients, such as that shown in Figure 3, the much faster response of the RGA is valuable to the operators to understand the transient behavior. This enables the operators to make smarter adjustments to the operation. The GC, in comparison, did not accurately capture all of the transient information recorded by the RGA.

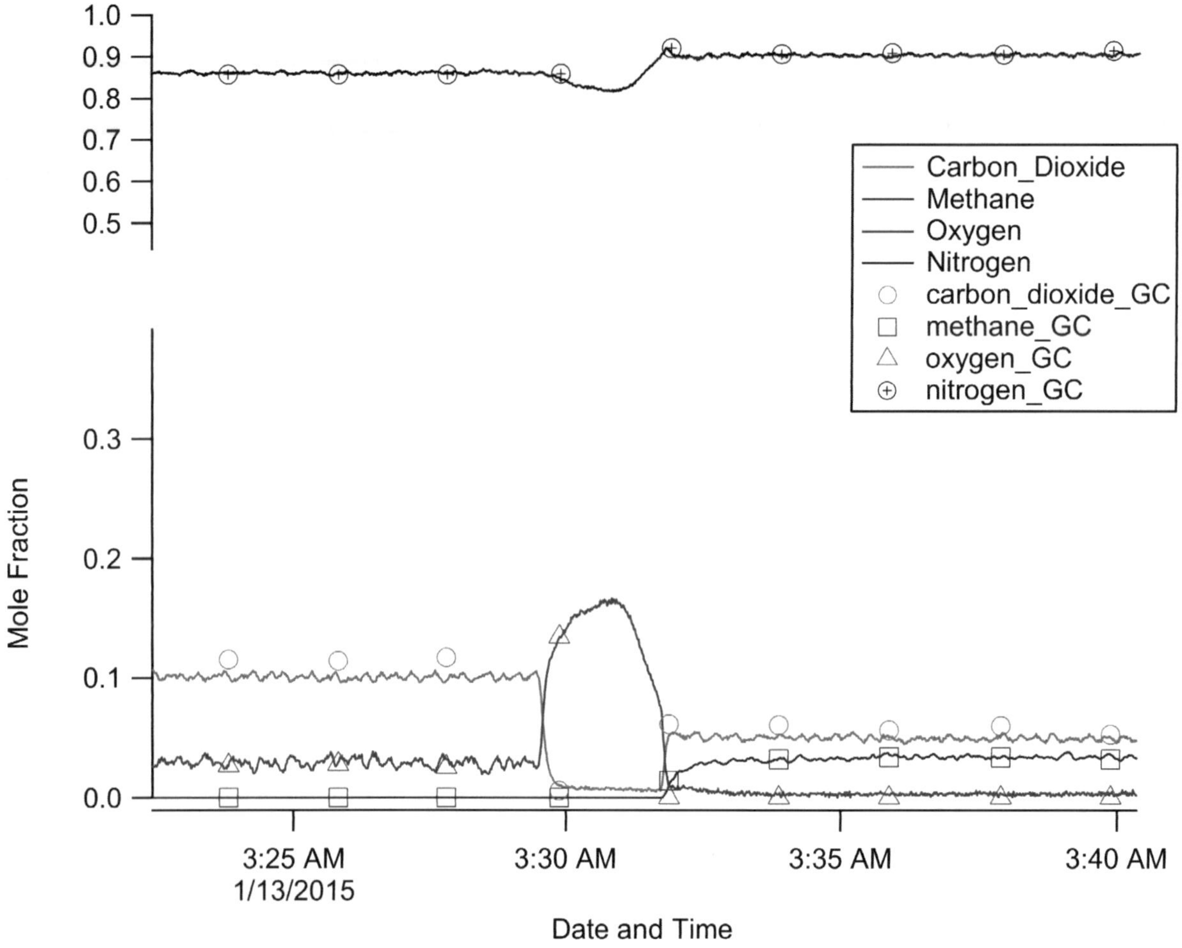

Figure 3. Raman Gas Analyzer (lines) and GC data (open symbols) from CLR operation, showing data during a rapid change.

Conclusions

The novel configuration of a reflectively lined capillary as a flow through sample cell and optical element has provided a large improvement in signal collection from gas phase Raman scattering, enabling the development of a new instrument with real-time gas analysis capability for major concentration species. NETL has performed the R&D to move this technology from the lab bench to a field application prototype. Data from monitoring an experimental chemical looping reactor is reported. Fast multi-species gas composition monitoring has been demonstrated and is particularly valuable for observation and control of transient events. For the chemical looping application, the clear difference in the spectra of carbon monoxide and nitrogen is a benefit of the Raman Gas Analyzer in comparison to a mass spectrometer. Real-time analysis of the gas composition allows operators to have feedback of process changes on the order of seconds as opposed to minutes with the GC, helping to improve control of the process.

Future plans include additional testing of the RGA in support of DOE research applications, support of technology transfer activities, and minor refinements to the field test units.

Acknowledgements

This project was funded by the Department of Energy, National Energy Technology Laboratory, an agency of the United States Government. Neither the United States Government nor any agency thereof, nor any of their employees, makes any warranty, expressed or implied, or assumes any legal liability or responsibility for the accuracy, completeness, or usefulness of any information, apparatus, product, or process disclosed, or represents that its use would not infringe privately owned rights. Reference herein to any specific commercial product, process, or service by trade name, trademark, manufacturer, or otherwise, does not necessarily constitute or imply its endorsement, recommendation, or favoring by the United States Government or any agency thereof. The views and opinions of authors expressed herein do not necessarily state or reflect those of the United States Government or any agency thereof.

References

1. Ekström et. al (2009), "Techno-Economic Evaluations and Benchmarking of Pre-combustion CO2 Capture and Oxy-fuel Processes Developed in the European ENCAP Project," Energy Procedia, Volume 1, Issue 1, pp. 4233-4240.
2. Alstom Power Inc., Power Plant Laboratories, "Greenhouse Gas Emissions Control by Oxygen Firing in Circulating Fluidized Bed Boilers: Phase 1 – A Preliminary Systems Evaluation," Vol 1., DE-FC26-01NT41146, http://www.netl.doe.gov/File%20Library/Research/Coal/ewr/co2/41146-Alstom_Power_Final-I_2003_oxycombustion-CFB_R01_Volume.pdf
3. Michael P. Buric, Benjamin T. Chorpening, Jessica C. Mullen, Joseph A. Ranalli, and Steven D. Woodruff, 2012, "Field testing the Raman gas composition sensor for gas turbine operation," SPIE Defense Sensing and Security Conference
4. Buric, M.P., Chen, K.P.,Falk, J., Woodruff, S.D., "Multimode metal-lined capillaries for Raman collection and sensing", JOSA B, Vol. 27, Issue 12, pp. 2612-2619 (2010)
5. http://www.netl.doe.gov/newsroom/labnotes/jan-2014
6. R. Breault, G. Richards, and S. Carpenter, "Update on NETL's Industrial Carbon Management Initiative: A Review of 2013 Activities," 2014 Clearwater Coal Conference, June 2-6, 2014
7. J. Weber, D. Straub, R. Breault, G. Richards, "Operating Experience of a CL Circulating Fluidized Bed Combustor," 2014 Clearwater Coal Conference, June 2-6, 2014

429981 Validation of Particle Fluid Dynamic Code Using Chemical Looping Experiments

Abstract:

Chemical looping combustion (CLC) and chemical looping with oxygen uncoupling (CLOU) are promising carbon capture technologies. Little experimental or simulated work has been copleted on systems larger than 1 MW. A complete understanding of the hydrodynamics will result in a quicker scaling to larger systems. To understand the hydrodynamics a scaled cold-flow unit was evaluated through visualization, pressure measurements, circulation rate, and bed mass changes. To further explore the hydrodynamics of CLC and CLOU a 10 kW system was run hot and cold was evaluated through pressure measurements, circulation rate, and bed mass changes. These evaluations and experiments were compared to simulations of Barracuda-VR®. The results shows that Barracuda-VR® accuratly represents the experiments.

Sensitivity of Chemical-Looping Combustion to Particle Reaction Kinetics

Matthias A. Schnellmann *(mas227@cam.ac.uk)*, Stuart A. Scott, G. Williams, John S. Dennis

AIChE Annual Meeting 2015

Salt Lake City, Thursday 12th Nov 2015

Outline

- Chemical-looping combustion concept
- Modelling
- Results
- Conclusions
- Future work

Chemical-Looping Combustion Concept

- Carbon capture technology
- Based on redox cycling of an oxygen carrier.
- Lattice oxygen in carrier reacts with fuel in fuel reactor (1)
- Gaseous product, after removal of H_2O, is a pure stream of CO_2.
- Reduced oxygen carrier is re-oxidised with air in air reactor (2)
- Inherent CO_2 separation achieved at fraction of energy penalty of other techniques
- Require stable, reactive and durable oxygen carriers

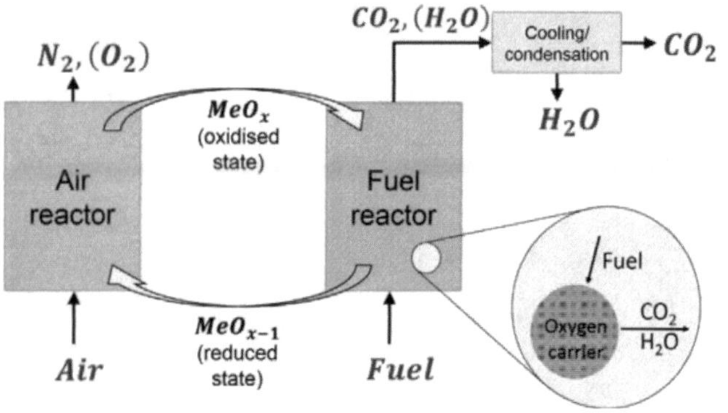

$$4MeO + CH_4 \rightarrow 4Me + 2H_2O + CO_2 \qquad (1)$$

$$4Me + 2O_2 \rightarrow 4MeO \qquad (2)$$

UNIVERSITY OF CAMBRIDGE

M.A. Schnellmann (mas227@cam.ac.uk) 3

Challenges being addressed

- Significant research on measuring redox kinetics of oxygen carriers at the laboratory scale.
- Can be almost impossible to determine *e.g.* rate of oxidation at high temperature.

How sensitive is CLC to the kinetics of individual particles?

What impact should this have when designing a CLC system?

What from smaller scales is important at larger scales?

UNIVERSITY OF CAMBRIDGE

M.A. Schnellmann (mas227@cam.ac.uk) 4

Model Development – Cycling behaviour

- Two, coupled fluidised bed reactors, with steady circulation of particles between them.
  - Well-mixed with respect to gas and solids
  - No reaction or mixing in the connecting lines
  - All particles have identical physical and chemical properties, except state of conversion
  - No elutriation
- Residence time distribution (RTD) of particles $$E(t) = \frac{1}{\bar{t}} e^{-\frac{t}{\bar{t}}}$$

$\bar{t}$ - mean residence time *t – actual time spent in reactor*

Model Development – Cycling behaviour

- Monte Carlo simulation used to determine the mean conversion and rate of reaction as particles leave the reactors
- I.C. – a fully oxidised particle
- Inverse Transform Sampling used to determine times from the RTD function:

$$t = -\bar{t}ln(U)$$

U – random number between 0 and 1

- Particle is cycled 1000 times
- Similar to method of Kimura *et al.* (1979)[1]
- Ergodic hypothesis

[1]Kimura, S., Fitzgerald, J., Levenspiel, O., & Kottman, C. (1979a). Solid Circulation Systems with Shrinking Core Kinetics in both Reactor and Regenerator. *Chemical Engineering Science, 34*, 1195–1201.

Particle models used

- Simple particle theories
  - Uniform reaction

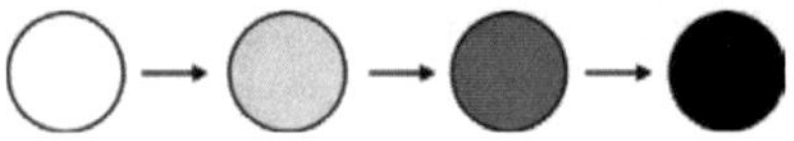

  - Shrinking Core Models (SCMs)

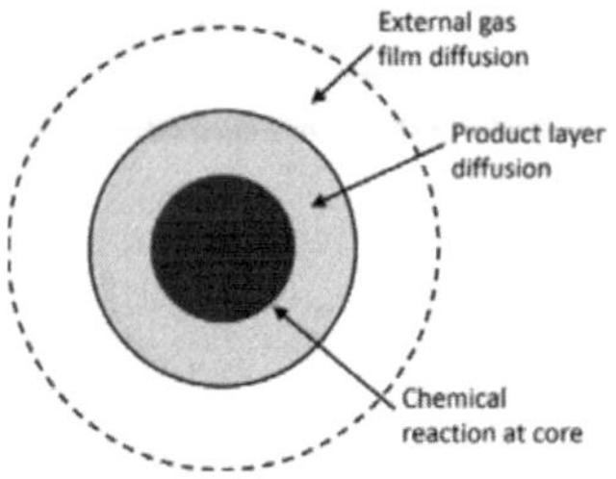

- Detailed particle model
  - Internal: Intrinsic kinetics coupled with Stefan-Maxwell equations
  - External: Stefan-Maxwell equations applied over a defined film thickness[1].
  - Energy equation
  - Spherical, homogeneous structure of monosized pores

[1]Hayhurst, A.N. (2000). The mass transfer coefficient for oxygen reacting with a carbon particle in a fluidized or packed bed. *Combustion and Flame*, *121*(4), 679-688. doi:10.1016/S0010-2180(99)00178-9

UNIVERSITY OF CAMBRIDGE

M.A. Schnellmann (mas227@cam.ac.uk)

Results

- Modelled $CuO:Al_2O_3$ (82.5 wt% CuO) oxygen carrier particle
- Temperature: 850°C
- Reduction: 5% CO, 2.5% CO_2, 92.5% N_2
- Oxidation: 5% O_2, 95% N_2

Reduction: $CuO + CO \rightarrow Cu + CO_2$ (1)

Oxidation: $2Cu + O_2 \rightarrow 2CuO$ (2)

Results

- Particle properties taken from the literature[1]

Property	Value
Diameter	300 μm
Skeletal density	4500 kg/m^3
Porosity	0.75
Tortuosity	$\sqrt{3}$
BET surface area	4.5 x 10^7 m^2/m^3
Mean pore radius	37 nm
Intrinsic first order rate constant (reduction)	6.5 x 10^{7}exp(-52 kJ/mol/RT) s^{-1}
Intrinsic first order rate constant (oxidation)	8.2 x 10^{6}exp(-44 kJ/mol/RT) s^{-1}

[1]Chuang, S. Y., Dennis, J. S., Hayhurst, A. N., & Scott, S. A. (2009). Kinetics of the chemical looping oxidation of CO by a co-precipitated mixture of CuO and Al2O3. *Proceedings of the Combustion Institute*, *32*(2), 2633–2640. doi:10.1016/j.proci.2008.06.112

UNIVERSITY OF CAMBRIDGE 9

Results – small number of cycles

- Using detailed particle model
- Particle is initially fully oxidised

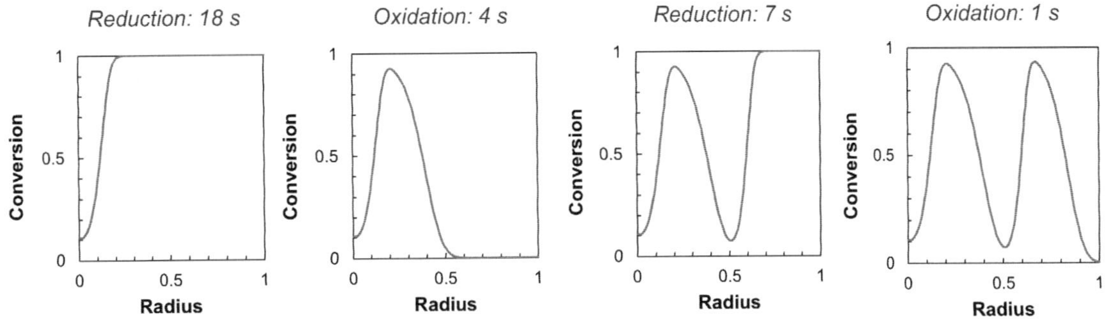

Conversion: 0 – oxidised, 1 – reduced
Radius: 0 – centre, 1 - surface

UNIVERSITY OF CAMBRIDGE *M.A. Schnellmann (mas227@cam.ac.uk)* 10

Results – Simple theories

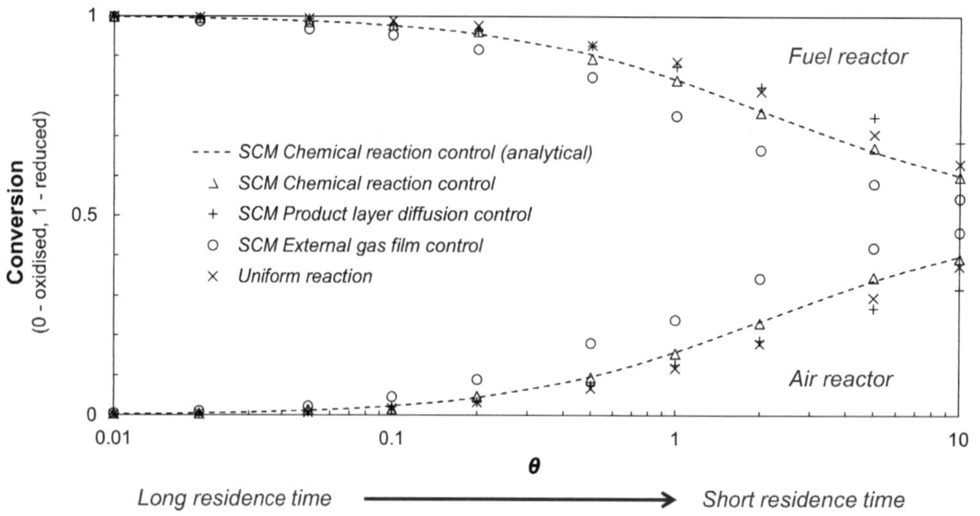

θ = total reaction time / mean residence time

Results – Simple theories

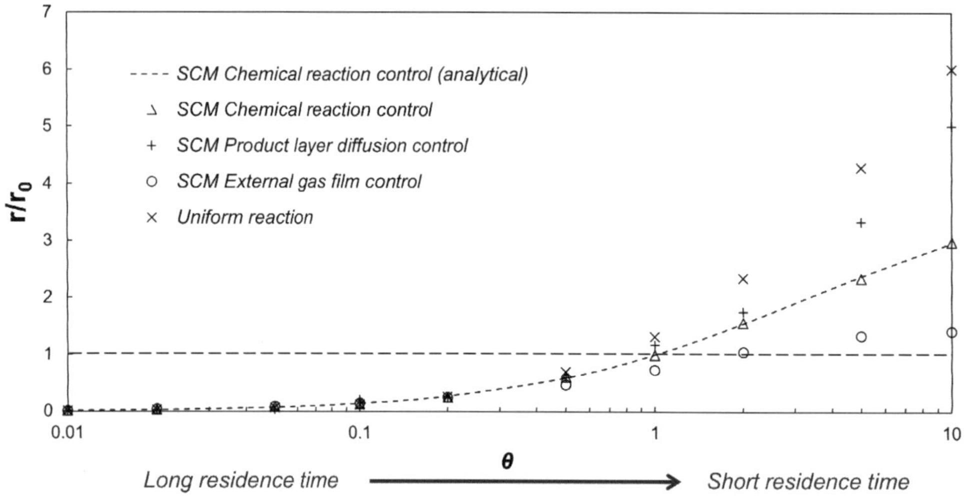

θ = total reaction time / mean residence time

R_0 is analytical value for a SCM under chemical reaction control when θ = 1.

Results – Detailed particle model

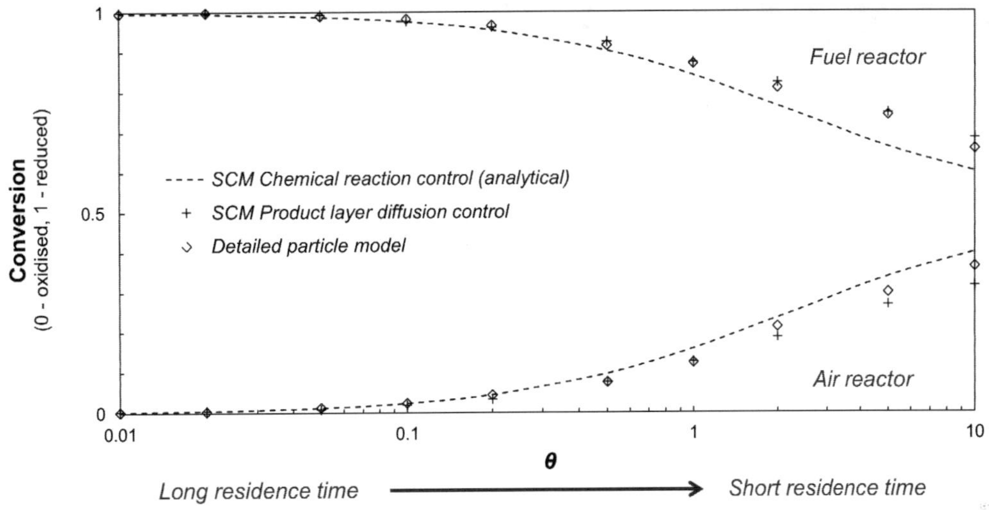

θ = total reaction time of particle / mean residence time of particle in reactor

UNIVERSITY OF CAMBRIDGE M.A. Schnellmann (mas227@cam.ac.uk) 13

Results – Detailed particle model

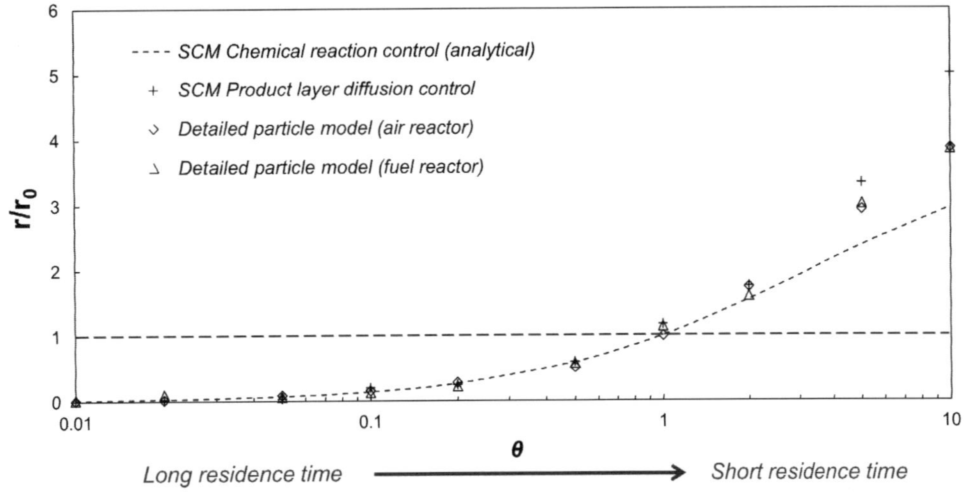

θ = total reaction time of particle / mean residence time of particle in reactor
R_0 is analytical value for a SCM under chemical reaction control when θ = 1.

UNIVERSITY OF CAMBRIDGE M.A. Schnellmann (mas227@cam.ac.uk) 14

Results – Different θ

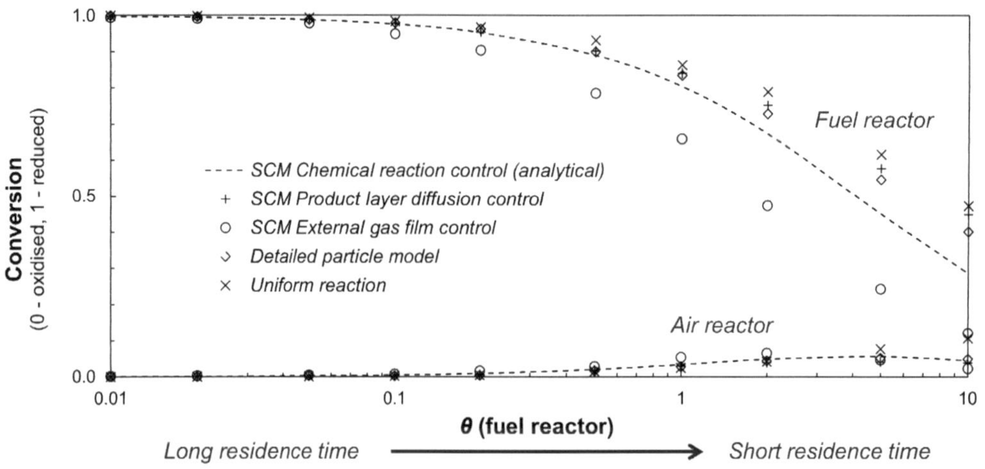

θ = total reaction time of particle / mean residence time of particle in reactor
θ in the air reactor is five times smaller than in the fuel reactor.

Results – Different θ

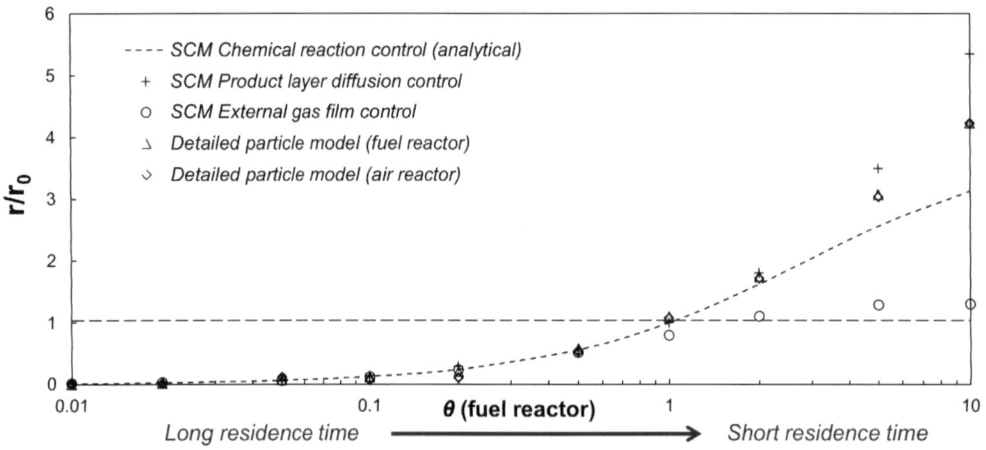

θ = total reaction time of particle / mean residence time of particle in reactor
R_0 is analytical value for a SCM under chemical reaction control when θ = 1.
θ in the air reactor is five times smaller than in the fuel reactor.

Conclusion

- Simple simulation for CLC has been developed
- Conversion history can be important for CLC
- CLC relatively insensitive to different particle kinetics when $\theta < 1$
- Also applicable to other reactor-regenerator systems
- Scope for using simple models of particle reaction to approximate the detailed behaviour in overall models of the CLC system
- For the detailed case modelled, a SCM under product layer diffusion control is suitable.

Future work

- Add a particle size distribution and investigate its impact
- Take account of how the particle evolves with time, both physically and chemically
- Develop a more detailed model of the CLC system

Acknowledgements

EPSRC

Johnson Matthey

Selwyn College, University of Cambridge

Questions?

Matthias A. Schnellmann
Department of Chemical Engineering and Biotechnology
University of Cambridge, UK
mas227@cam.ac.uk

Mathematical Modeling of Redox Reactions of Copper-Based Oxygen Carriers with Sintering Effects

Juan C. Maya and Farid Chejne*

fchejne@unal.edu.co

NCOMBUSTIO
UNIÓN TEMPORAL

TRABAJO Y RECTITUD

UNIVERSIDAD NACIONAL DE COLOMBIA
SEDE MEDELLÍN
FACULTAD DE MINAS

Schedule

1. Introduction
2. The Novel Model and Solution Strategy
3. Experimental procedure
4. Results and Discussion

TRABAJO Y RECTITUD

UNIVERSIDAD NACIONAL DE COLOMBIA
SEDE MEDELLÍN
FACULTAD DE MINAS

1. Introduction

TRABAJO Y RECTITUD

UNIVERSIDAD NACIONAL DE COLOMBIA SEDE MEDELLÍN FACULTAD DE MINAS

Changing Grain Size Model: Georgakis, Chang, and Szekely 1979

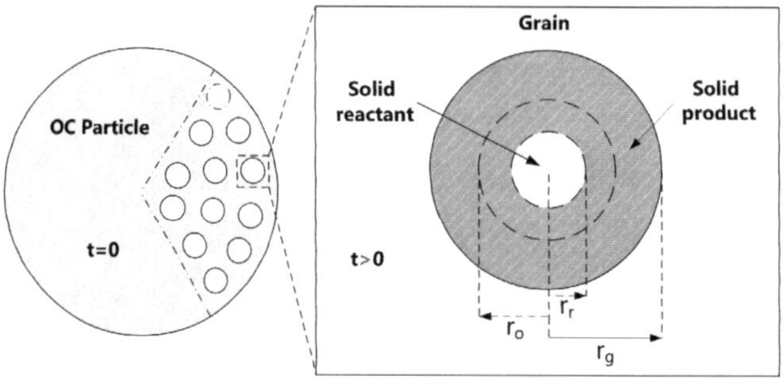

Changing Grain Size Model applied to CLC: Abad et al., 2007; Celaya, 2007; García-Labiano, de Diego, Adánez, Abad, & Gayán, 2005; Song et al., 2008.

TRABAJO Y RECTITUD

UNIVERSIDAD NACIONAL DE COLOMBIA SEDE MEDELLÍN FACULTAD DE MINAS

Shortcomings of grain models:

- Grains with initial uniform radius are assumed
- Sintering effects are not taken into account.

TRABAJO Y RECTITUD

UNIVERSIDAD NACIONAL DE COLOMBIA SEDE MEDELLÍN FACULTAD DE MINAS

Some approaches :

Distributed, no sintering

- Population balance grain model (Bhattacharya & Purohit, 2004).
- Distributed grain model for redox reactions of Cu-based-OC (Maya & Chejne, 2014)

Sintering, non-distributed

- Grain model with sintering represented as combination of grains for ZnO reactions (Manovic, Grubor, & Loncarevic, 2006; Ranade & Harrison, 1979)

TRABAJO Y RECTITUD

UNIVERSIDAD NACIONAL DE COLOMBIA SEDE MEDELLÍN FACULTAD DE MINAS

2. The New Model

TRABAJO Y RECTITUD

UNIVERSIDAD NACIONAL DE COLOMBIA
SEDE MEDELLÍN
FACULTAD DE MINAS

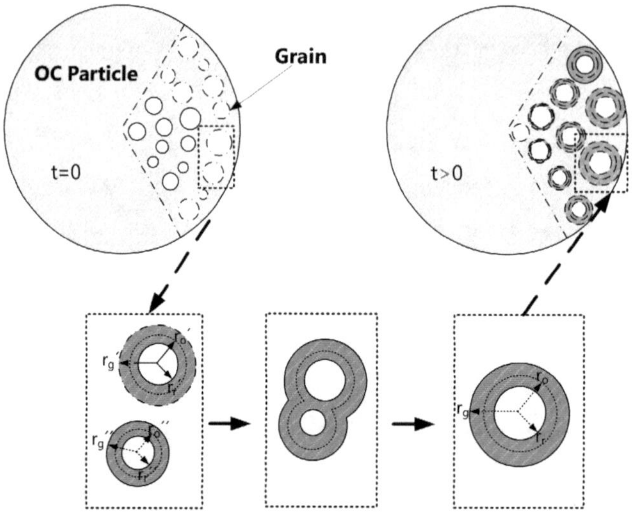

TRABAJO Y RECTITUD

UNIVERSIDAD NACIONAL DE COLOMBIA
SEDE MEDELLÍN
FACULTAD DE MINAS

Model Assumptions:

- The pellet is spherical and made of nonporous spherical grains.
- The pellet retains its spherical shape and its diameter during the reaction.
- At the beginning of reaction, the grain sizes are distributed non-uniformly.
- Sintering is modeled as a phenomenon of combination of grains.
- Isothermal process.

TRABAJO Y RECTITUD

UNIVERSIDAD NACIONAL DE COLOMBIA SEDE MEDELLÍN FACULTAD DE MINAS

$H_2 + CuO \rightarrow H_2O + Cu$ $\qquad$ ${}^{1}/_{2}\, O_2 + Cu \rightarrow CuO$

$$S(solid) + A(gas) \rightarrow C(solid) + D(gas)$$

Particle mass balance:

$$\varepsilon \frac{\partial C_A}{\partial t} = \frac{1}{R^2}\frac{\partial}{\partial R}\left(D_e R^2 \frac{\partial C_A}{\partial R}\right) + \underbrace{\rho_{m,S}\int_0^{\infty} \frac{dr_r}{dt} 4\pi {r_r}^2 f_r dr_r}_{Connection\ term}$$

Subject to

$\frac{\partial C_A}{\partial R} = 0$ $\qquad$ R=0

$C = C_{A,b}$ $\qquad$ $R=R_p$

TRABAJO Y RECTITUD

UNIVERSIDAD NACIONAL DE COLOMBIA SEDE MEDELLÍN FACULTAD DE MINAS

Chemical Reaction

$$\frac{dr_r}{dt} = -\frac{kC_A}{\rho_{m,S}\left[1+\frac{k}{D_S}r_r\left(1-\frac{r_r}{r_g}\right)\right]}$$

$$\frac{dr_g}{dt} = (1-Z)\left(\frac{{r_r}^2}{{r_g}^2}\right)\left(\frac{dr_r}{dt}\right)$$

$$Z = \frac{\rho_{m,S}}{\rho_{m,C}}$$

TRABAJO Y RECTITUD

UNIVERSIDAD NACIONAL DE COLOMBIA SEDE MEDELLÍN FACULTAD DE MINAS

Population Balance:

$$\frac{\partial f_j}{\partial t} + \underbrace{\frac{\partial}{\partial r_j}\left[\left(\frac{\partial r_j}{\partial t}\right) f_j\right]}_{Reaction} - \underbrace{B + D}_{Sintering} = 0$$

Where $j = o, r, g$

Aggregation Frequency

$$\beta = \varphi\hat{\beta}$$

$$\hat{\beta} = \frac{1}{\tau_{coag}\frac{1}{2}\mu_{j,0}(t)}$$

TRABAJO Y RECTITUD

UNIVERSIDAD NACIONAL DE COLOMBIA SEDE MEDELLÍN FACULTAD DE MINAS

3. Experimental Procedure

TRABAJO Y RECTITUD

UNIVERSIDAD NACIONAL DE COLOMBIA SEDE MEDELLÍN FACULTAD DE MINAS

Thermogravimetry

T=1073 K
Particle size~500 µm

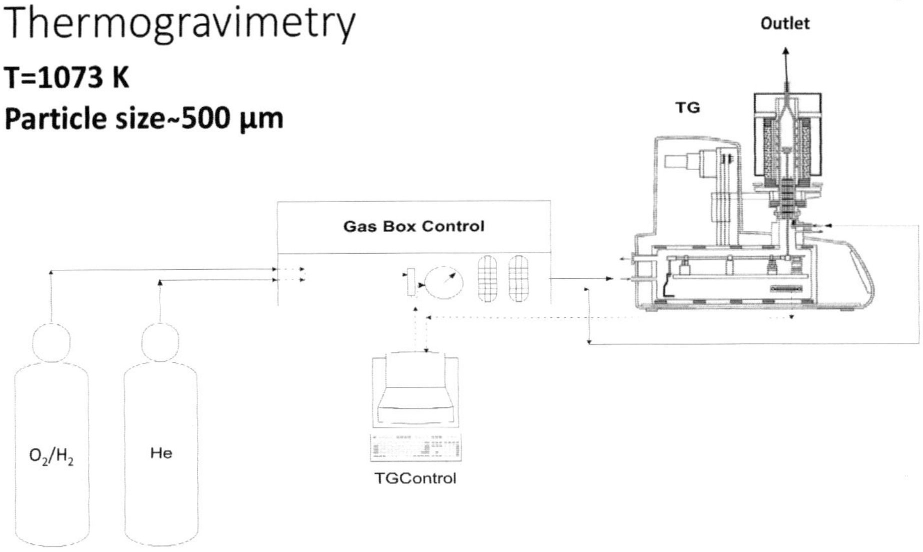

Image taken from **Valdés et al. (2015)**

TRABAJO Y RECTITUD

UNIVERSIDAD NACIONAL DE COLOMBIA SEDE MEDELLÍN FACULTAD DE MINAS

4. Results and Discussion

Oxidation Cycle 1, I20

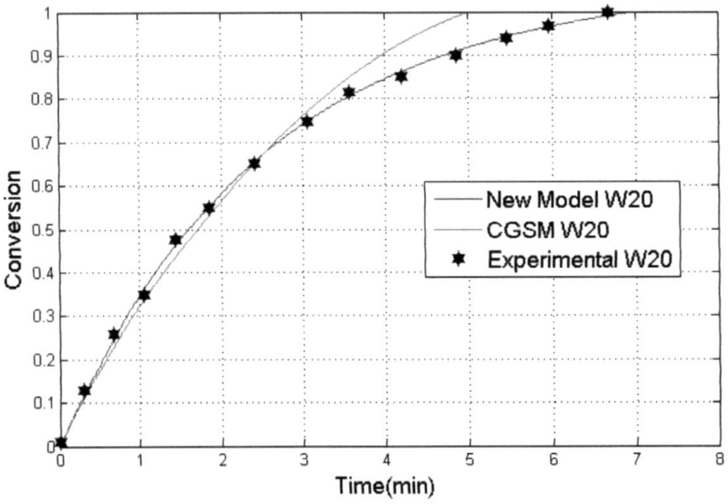

The authors wish to thank the "Departamento Administrativo de Ciencia, Tecnología e Innovación" - COLCIENCIAS (Administrative Department of Science, Technology and Innovation from Colombia) through the program "Research on advanced combustion innovation in Industrial use", code N° 1115-543-31906 contract N° 0852-2012.

THANK YOU

TRABAJO Y RECTITUD

UN
UNIVERSIDAD NACIONAL DE COLOMBIA
SEDE MEDELLÍN
FACULTAD DE MINAS

434116 Dynamic Simulation of a Natural Gas-Fired Combined Cycle Power Plant Integrated with Chemical-Looping Combustion

Abstract:

Dynamic Simulation of a Natural Gas-Fired Combined Cycle Power Plant Integrated with Chemical-looping Combustion

Lu Han, Chen Chen, Kyle D. Such, Zhiquan Zhou, **George M. Bollas**[†]

Department of Chemical & Biomolecular Engineering, University of Connecticut, Storrs, CT

Abstract

Chemical-looping combustion (CLC) offers a promising method to reduce the CO_2 emissions from power generation, at minimal energy penalty.[1] The fixed-bed reactor is recognized as a promising reactor configuration for CLC because it avoids the problems with gas/solid separation and particle attrition commonly reported in circulating processes.[2] Also, the fixed-bed is more compact and easier to scale-up and pressurize. The major disadvantage of the fixed-bed reactor is its dynamic operation, leading to fluctuations in temperature and flow rate of the streams exiting the reactor. This is challenging when the CLC reactor is integrated into a gas turbine-based combine cycle, because the downstream gas turbine is most efficient when operated at the nominal steady-state conditions.

In this work, we evaluate the dynamic operation of fixed-bed reactors and their impact on the power plant performance. Nickel- and copper-based oxygen carriers are exploited for their high reactivity and stability in the CLC cycles.[3,4] First, a detailed kinetic model is needed to capture the reactivity of the oxygen carriers at the high-pressure operation of modern power cycles. Here, the kinetics of the oxygen carrier reduction and oxidation reactions are obtained from high pressure experiments conducted in a bench-scale fixed-bed reactor under methane and syngas fuel and various pressures from 1 to 10 bar.[5] An empirical model[6] is used to describe the effect of pressure on the kinetic rates, leading to accurate predictions of the data. Subsequently, the design of the CLC reactor is treated as a dynamic optimization problem, in order to maximize the efficiency of the power plant. The temperature fluctuations of the reactor exhaust are minimized to satisfy the standard operating conditions of commercial gas turbines, by manipulating the mass flows to the reactor, cycle times, and active metal loading of the oxygen carriers. In addition, the CLC reactor must exhibit a high fuel conversion (>98%), sufficient CO_2 capture efficiency (>90%), and low pressure drop (<8%), which are set as constraints within the optimization problem.

We propose optimal designs for the CLC reactor utilizing nickel and copper oxygen carriers and explore the effect of the residual dynamics on the performance of the combined cycle power plant. A power plant model is developed in Dymola,[7] consisting of a gas turbine and a steam cycle. The components of the model are separately validated with steady-state data from an existing plant.[8] The system variables predicted by the model are within a narrow range of the reported values, and therefore the models can be used for dynamic analysis. In this work, the exhaust streams of the CLC reactor are fed to the gas turbine and bottoming cycles, to simulate an integrated natural-gas fired power plant with CLC. The feasibility of the combined cycle power plant to reach continuous operation is demonstrated, despite the inherent dynamics of the cyclic operation of the alternating-flow CLC reactor. By applying this model-based approach to the CLC power plant design, we are more equipped to examine the real efficiencies of the plant and propose methods to increase the heat recovery cycle for higher efficiencies.

Acknowledgement

This material is based upon work supported by the National Science Foundation under Grant No. 1054718. Financial support by Alstom Power, Inc. is gratefully acknowledged.

References

(1) Adanez, J.; Abad, A.; Garcia-Labiano, F.; Gayan, P.; De Diego, L. F. Progress in Chemical-Looping Combustion and Reforming Technologies. *Prog. Energy Combust. Sci.* **2012**, *38* (2), 215.

(2) Noorman, S.; van Sint Annaland, M. Packed Bed Reactor Technology for Chemical-Looping Combustion. *Ind. Eng. Chem. Res.* **2007**, *46* (12), 4212.

(3) Mattisson, T.; Johansson, M.; Lyngfelt, A. The Use of NiO as an Oxygen Carrier in Chemical-Looping Combustion. *Fuel* **2006**, *85* (5-6), 736.

(4) García-Labiano, F.; de Diego, L. F.; Adánez, J.; Abad, A.; Gayán, P.; Garcia-Labiano, F.; Diego, L. De. Reduction and Oxidation Kinetics of a Copper-Based Oxygen Carrier Prepared by Impregnation for Chemical-Looping Combustion. *Ind. Eng. Chem. Res.* **2004**, *43* (26), 8168.

(5) Nordness, O.; Zhou, Z.; Bollas, G. M. Feasibility of Integration of Chemical Looping Combustion (CLC) in IGCC Systems: High Pressure CLC Experiments with CH_4 and Syngas. In *AIChE Annual Meeting*; Atlanta, GA, 2014.

(6) García-Labiano, F.; Adánez, J.; de Diego, L. F.; Gayán, P.; Abad, A.; Cuadrat, A. Effect of Pressure on the Behavior of Copper-, Iron-, and Nickel-Based Oxygen Carriers for Chemical-Looping Combustion. *Energy & Fuels* **2005**, *24* (10), 1402.

(7) Cellier, F. E. Dymola: Environment for Object-Oriented Modeling of Physical Systems, 2015.

(8) Singer, J. G. *Combustion Fossil Power: A Reference Book on Fuel Burning and Steam Generation*; Combustion Engineering, 1991.

430100 Coal to Methanol Process Using Chemical Looping Gasification for Syngas Production and in-Situ CO2 Capture: Experimental, Thermodynamic and Techno-Economic Analysis

Abstract:

Methanol is an important readily-transportable intermediate which is used to produce multiple chemical products including olefins, gasoline, dimethyl ether (DME) etc. Methanol is mainly produced from natural gas, with coal based production accounting for 9% of the total world demand. The use of coal gasification to produce methanol allows for the use of domestic coal in chemical manufacturing, which provides an opportunity for producing these commodities in the United States to provide supply security and an opportunity to export to overseas markets. While coal costs tend to be low, the capital costs for gasification are high as a result of the gasification equipment and air separation unit (ASU) used for oxygen production. Further, the overall yield of a coal to methanol process is low, leading to higher costs than a natural gas based process. The Ohio State (OSU) chemical looping gasification (CLG) technology for coal gasification to methanol production application uses an optimized iron-based oxygen carrier composite particle and a co-current downward reducer reactor configuration.

The optimized iron-based OSU oxygen carrier in-conjunction with a co-current moving bed reactor significantly increases the carbon conversion efficiency for syngas production and has been experimentally validated using bench-scale tests. WorleyParons have completed a techno-economic analysis which showed that for a coal only feed with 90% carbon capture, the OSU CLG technology reduced the methanol required selling price by 21%, lowered the capital costs by 28% and increased coal consumption efficiency by 14% over the reference case. Further, using the Ohio State Chemical Looping Gasification technology resulted in a methanol required selling price which was lower than the reference non-capture case. This presentation will initially focus on the theoritical thermodynamic rationale and validating experimental results for using a co-current moving bed reducer and an optimized oxygen carrier composite. The comprehensive techno-economic analysis performed in collaboration with WorleyParsons and current research for de-risking technology gaps for a successful commercial demonstration will be discussed.

426730 Determination of Autothermal State for Chemical Looping Combustion and Chemical Looping with Oxygen Uncoupling with Powder River Basin Coal As Fuel

Abstract:

Chemical Looping Combustion (CLC) and Chemical Looping with Oxygen Uncoupling (CLOU) are carbon-capture technologies that utilize a metal oxide as an oxygen carrier to selectively separate oxygen from air, for combustion. These technologies typically accomplish this separation using two reactors: an air reactor (AR) where the oxidation of metal is performed using ambient air, and a fuel reactor (FR) where the reduction of metal is accomplished and the combustion of fuels occurs. For optimal CLC and CLOU operation both reactors need to be exothermic to maintain the reactor temperature. For CLC and CLOU the oxidation of metal particles is an exothermic reaction resulting in an exothermic AR.

In CLC, gaseous fuel reacts with the metal oxide directly to form CO_2 and H

slightly endothermic. When utilizing a solid fuel in CLC the fuel needs to be gasified before reacting the fuel with the metal oxide. Gasification is a highly endothermic reaction which results in an overall endothermic FR. To run the FR as an exothermic reactor, energy needs to be provided. The optimal way to provide energy is through the hot oxygen carrier particles. To achieve this energy transfer the air reactor needs to operate at a higher temperature then the fuel reactor.

In CLOU, the metal oxides reduce by spontanously releasing oxygen which is directly used in the combustion of fuel. The reduction of metal oxides is an endothermic reaction, but the combustion of fuel with oxygen is highly exothermic. These reactions can result in an overall exothermic reaction, with both reactors being exothermic.
Using ASPEN Plus process modeling, temperature ranges were explored for when both reactors can be run exothermically. The process models took into account particle transfering energy, preheating of the air and steam, recycling flue gas, oxidation/reduction reaction, and the combustion of the fuel. ΔT is the temperature the AR is above the FR. For CLC, the FR temperature was 975 ^{o}C and it was 950 ^{o}C for CLOU. It was determined that in CLC the autothermal temperature range is $\Delta T = 75 - 125$ ^{o}C. In CLOU the autothermal temperature range is $\Delta T = 0 - 25$ ^{o}C.

437890 Integrating CO_2 Capture with Steam-Reforming Reactions

Abstract:

CO_2 capture by carbonation is an exothermic reaction, while CO_2 generation by steam-reforming of hydrocarbons or oils is endothermic. Capturing CO_2 under the reforming conditions does not only shift the thermodynamic equilibrium toward completion but also mitigates the heat transfer issue. In this presentation, research results reported in the literature will be reviewed to delineate the progress and remaining challenges. The recent development of novel sorbents suitable for in-situ capture applications will be reported, and its application to hydrogen production from bio-oil will be discussed.

2015 AIChE Annual Meeting (Salt Lake City)

An Analysis of Extraction Mechanism and the Role of Grinding in Carbon Dioxide Mineralization Processes

Azadeh Hemmati[a,b], Jalal Shayegan[b], Tze Yuen Yeo[a], Hamidreza Kariminia, Paul Sharratt[a], Jie Bu[a*]

bu_jie@ices.a-star.edu.sg

[a] Institute of Chemical and Engineering Sciences (ICES)
A*STAR, Singapore

[b] Sharif University of Technology (SUT)
Tehran, Iran

Contents

- **Why CO_2 Mineralization by Acid Extraction**
- **CO_2 Sequestration Efficiency and Energy Penalty of Grinding**
- **Mechanism of Mineral Acid Dissolution**
- **Conclusions**

CO_2 Mineralization Acid Extraction Process: Theoretical Reactions

Mineralization →

Recycle: $2\,NaCl$ (from the last reaction) → first reaction

$$NaCl + H_2O \longrightarrow NaOH + HCl$$

$$2\,NaOH + CO_2 \longrightarrow Na_2CO_3 + H_2O$$

$$Mg_3Si_2O_5(OH)_4 + 6\,HCl \longrightarrow 3\,MgCl_2 + 2\,SiO_2 + 5\,H_2O$$

$$MgCl_2 + Na_2CO_3 \longrightarrow MgCO_3 + 2\,NaCl$$

Net reaction: $Mg_3Si_2O_5(OH)_4 + 3CO_2 \rightarrow 3\,MgCO_3 + 2SiO_2 + 2H_2O$

Purification →

$$Fe^{3+}(aq) + 3OH^{-}(aq) \leftrightarrow Fe(OH)_3(s) \qquad (P_1)$$

$$Fe^{2+}(aq) + 2OH^{-}(aq) \leftrightarrow Fe(OH)_2(s) \qquad (P_2)$$

Hemmati, A., et al., *Solid products characterization in a multi-step mineralization process.* Chemical Engineering Journal, 2014. **252**: p. 210-219.

3

Mineralization with pH-swing Process

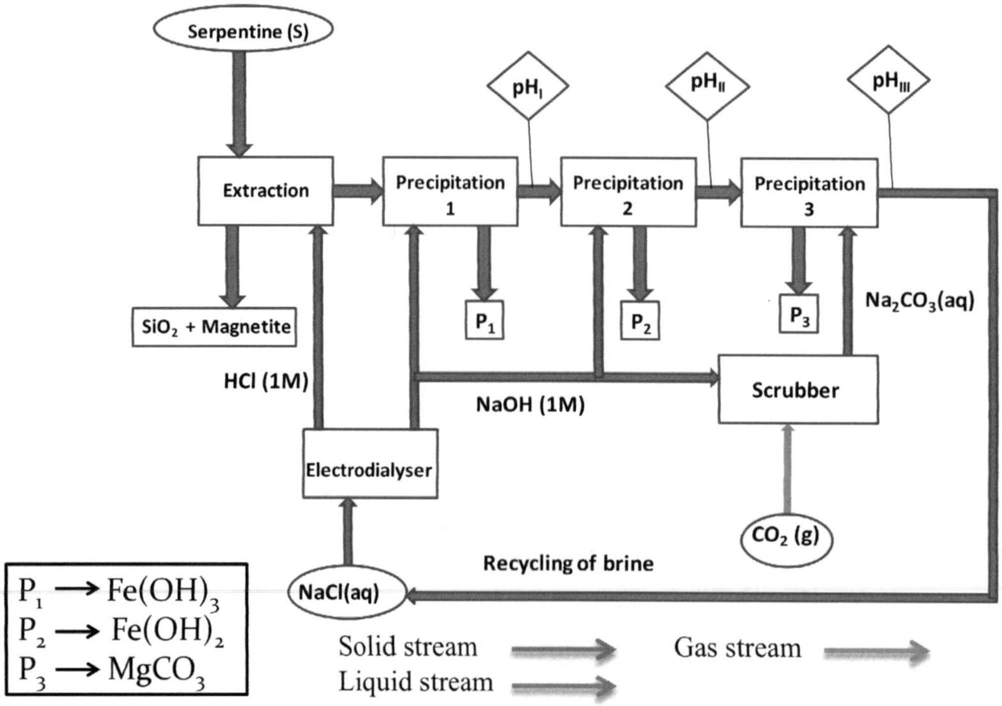

4

Materials

MgO	46.72%
SiO_2	42.70%
Fe_2O_3	8.40%

Raw materials
Serpentine
NaCl
CO_2
Water

Products
Silica
Magnetite
Iron hydroxide (II, III)
Hydrated $MgCO_3$

Indirect mineralization

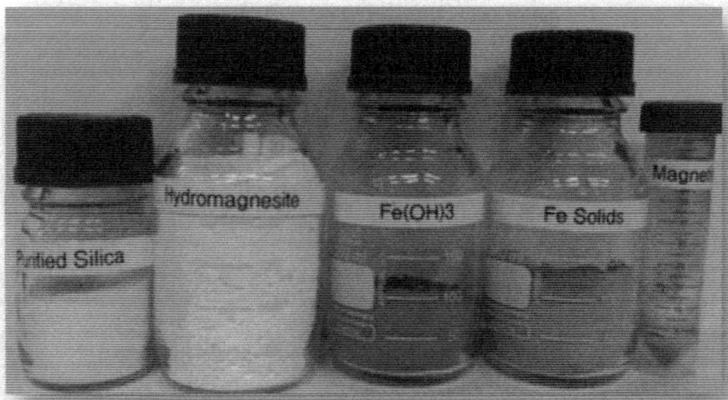

5

Magnesium Extraction

Serpentine crystals change to amorphous silica

Silica composition after 6 hr extraction reaction

SiO_2	MgO	Fe_2O_3	Cr_2O_3	Al_2O_3	Others
82.87%	14.91%	1.33%	0.43%	0.43%	0.03%

6

Serpentine Solid Composition After Extraction

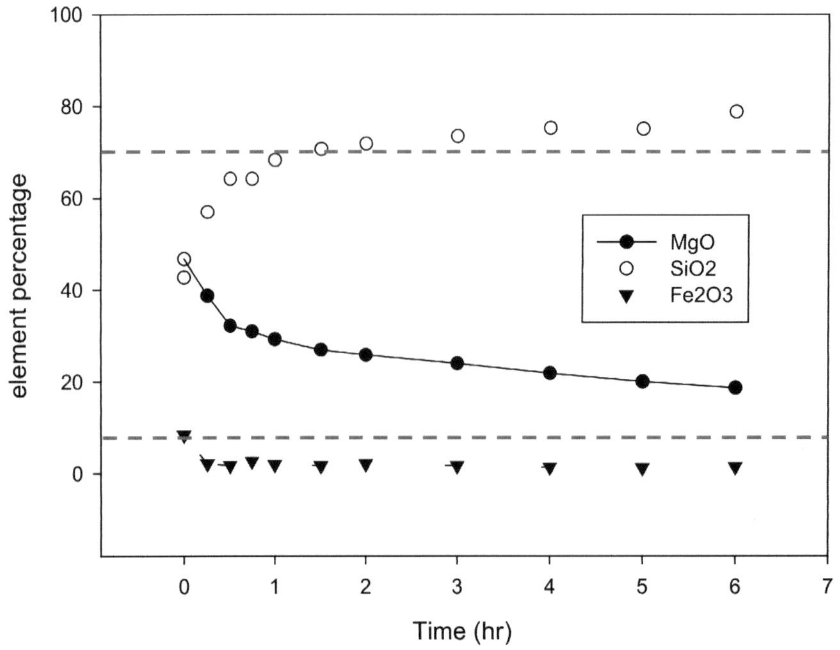

7

Particle Size Optimization

	D_{50} (μm) of mineral solid	Time	Stirring	Temp
Scenario 1	1,425	6 h	600rpm	80 °C
Scenario 2	625	6 h	600rpm	80 °C
Scenario 3	325	6 h	600rpm	80 °C
Scenario 4	187.5	6 h	600rpm	80 °C
Scenario 5	90.5	6 h	600rpm	80 °C
Scenario 6	30	6 h	600rpm	80 °C

8

Size Optimization for Serpentine

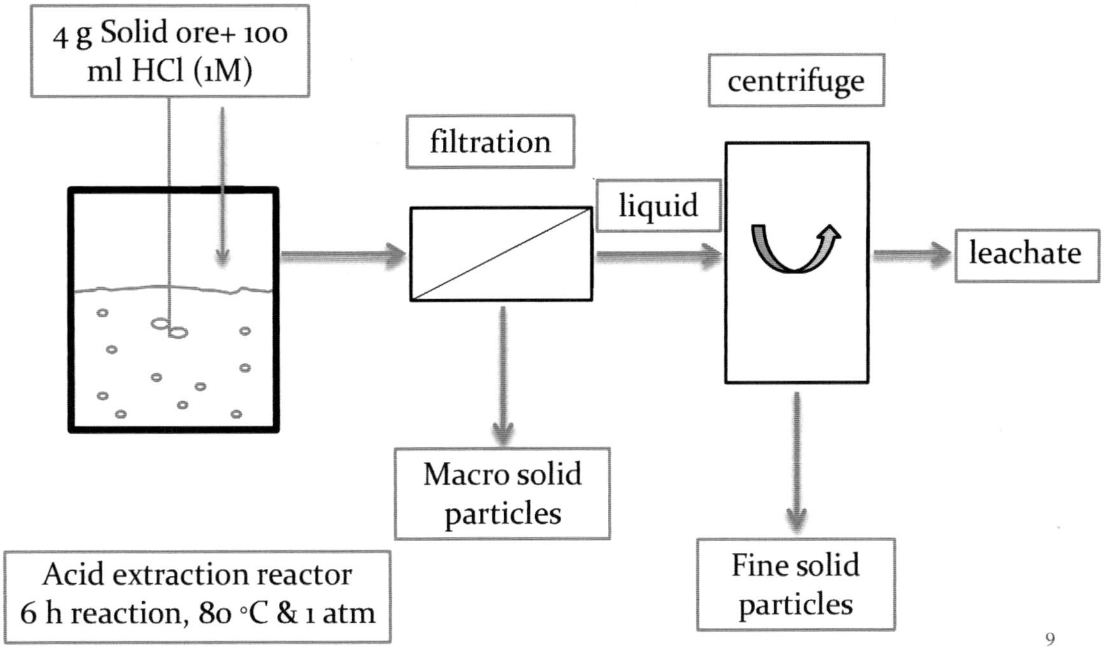

Mg-Extraction in Relation to Mineral Particle Sizes

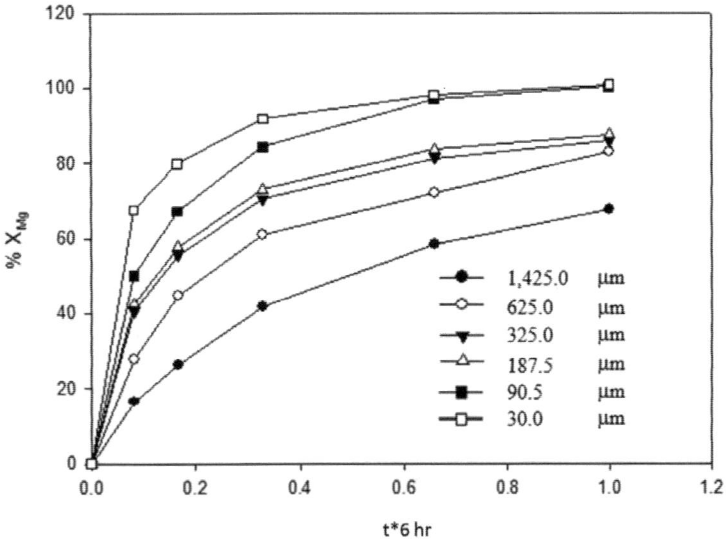

Mineral Grinding Energy

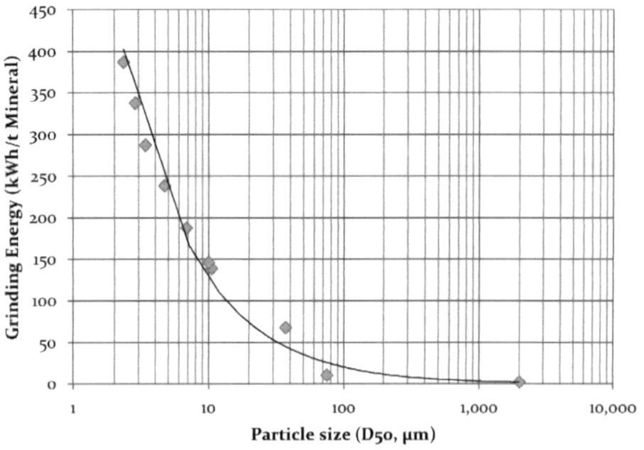

PD 50 (μm)	W_f (kWh/t mineral)
1425	4.2
625	6.55
325	8.54
187.5	12.47
90.5	19.1
30	33.74

•O'connor, W. et al. Aqueous mineral carbonation: Mineral availability, pretreatment, reaction parametrics, and process studies. A report produced by the Albany Research Centre for the US Department of Energy 2005.
•Gerdemann, S. et al. Environmental science & technology 2007, 41, (7), 2587-2593.
•Energy consumption for grinding based on bond's law (Particles size: 50mm > d > 0.05mm)

$$W_B = 10 \times C_B \left(\frac{1}{\sqrt{d_E}} - \frac{1}{\sqrt{d_A}}\right)$$

11

CO_2 Sequestration and Energy Penalty

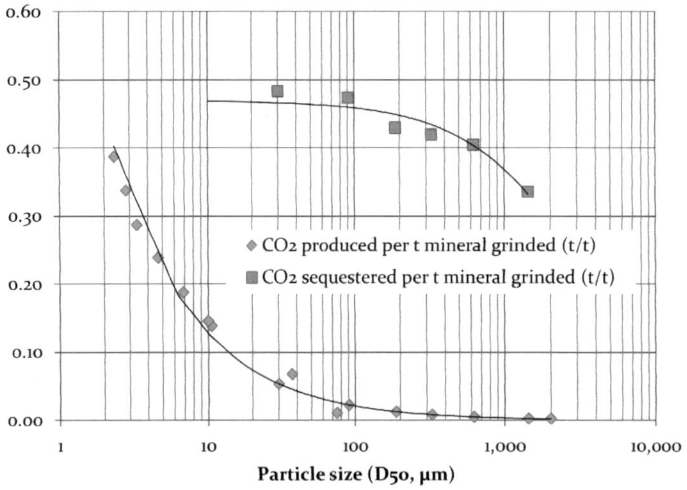

12

CO_2 Sequestration Efficiency and Energy Penalty

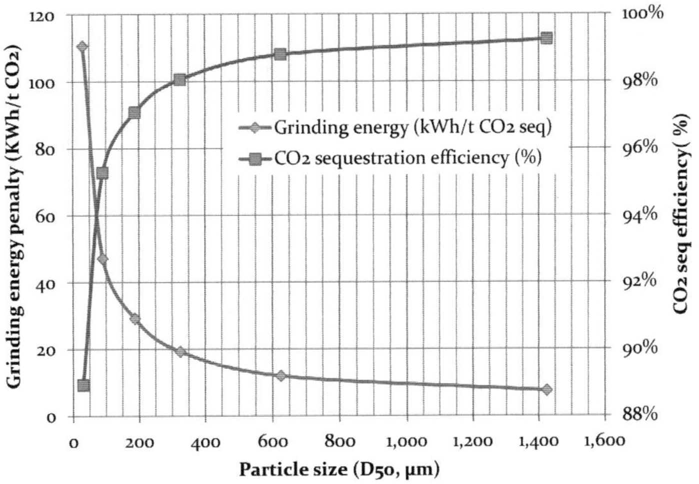

$$CO_2 Sequestration\ Efficiency = \frac{CO_2\ Sequestered - CO_2\ Emitted\ (Grinding)}{CO_2\ Sequestered}$$

13

Extraction Mechanism Investigation

Scenario	Stone Shape	Dimensions (cm)	Mass(g)	Surface removal	Agitation	Reaction time (h)
1	Sphere	R=0.694	3.6258	No	Yes (600 rpm)	90
2	Cube	2×1.2×0.8	6.9494	No	No	60
3	Cube	2×1.1×0.9	6.6307	Yes	No	60

Rate of Mass Decrease per Surface Area vs Time (Scenario 1)

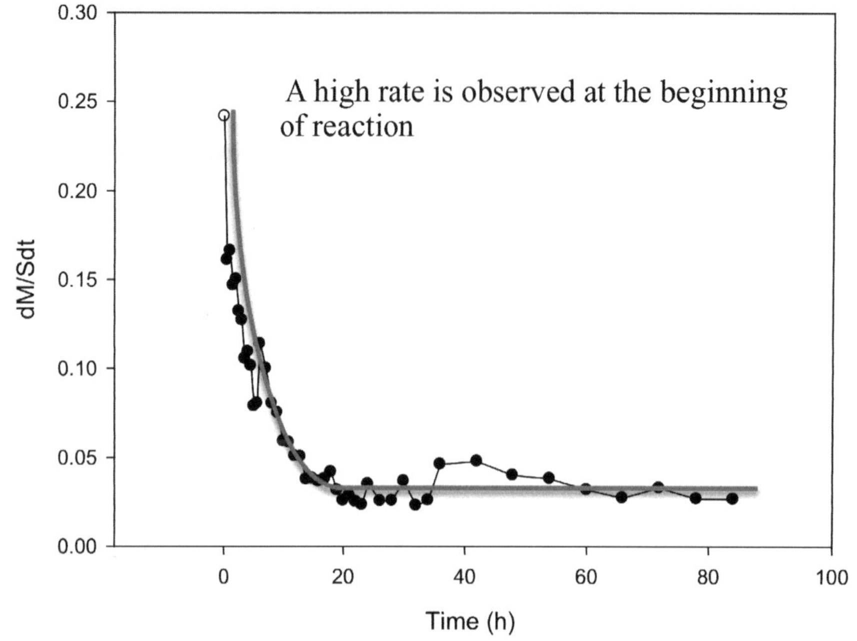

15

Extracted Mg/Extracted Fe (Scenario 1)

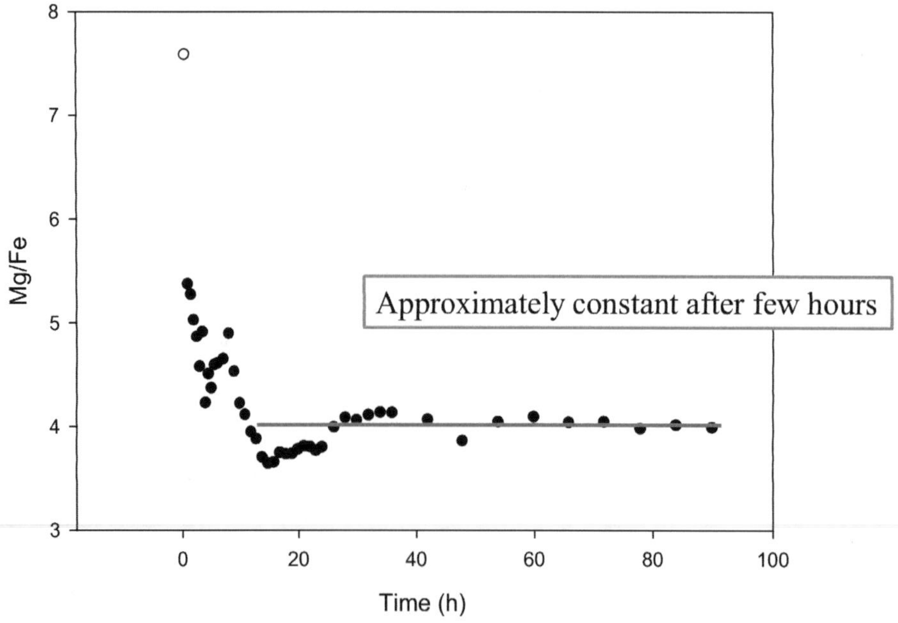

16

Study on Macro Particles: Scenario 2

6.9494 g sphere mineral + 100 ml HCl (1M)

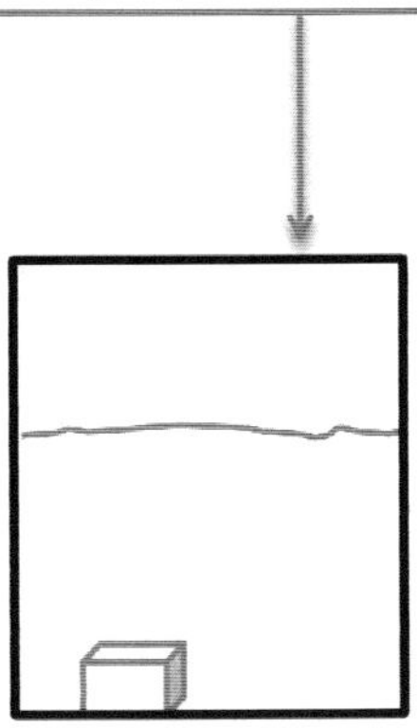

Acid dissolution reactor
60 h reaction at 80 °C & 1 atm

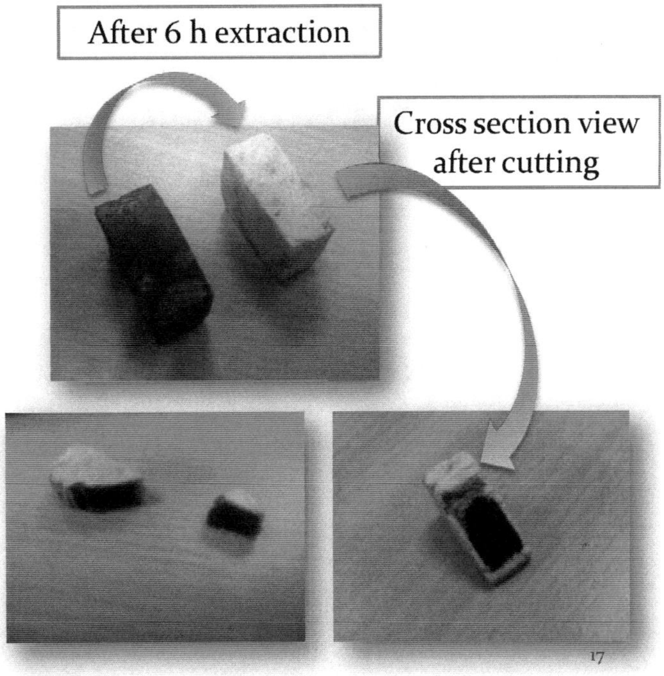

Reduction in Particle Mass (Scenario 2)

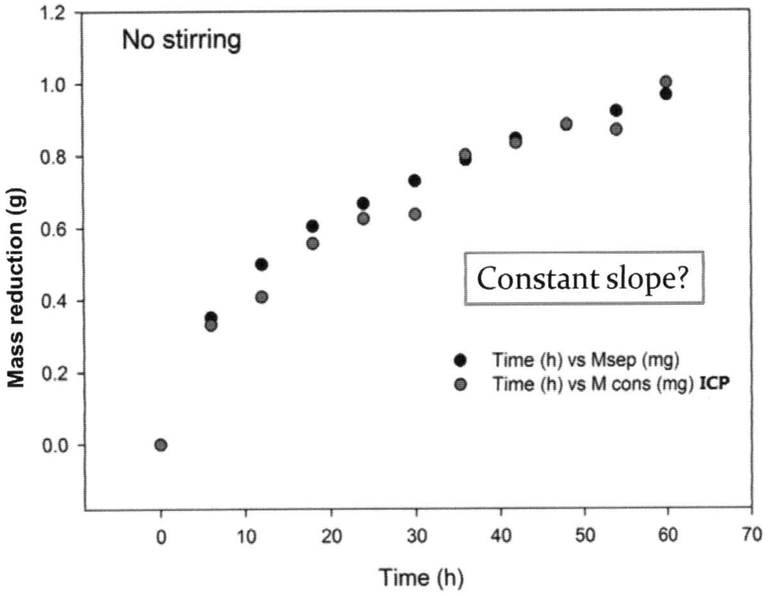

18

Study on Macro Particle: Scenario 3 Regular Removal of Surface Silica Layer

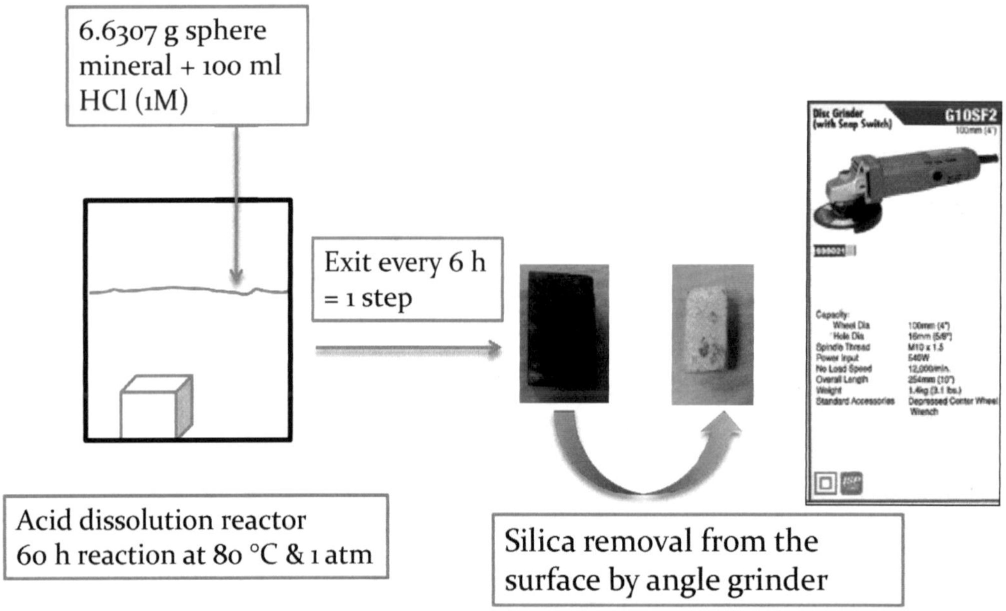

19

Reduction in Particle Mass (All Scenarios)

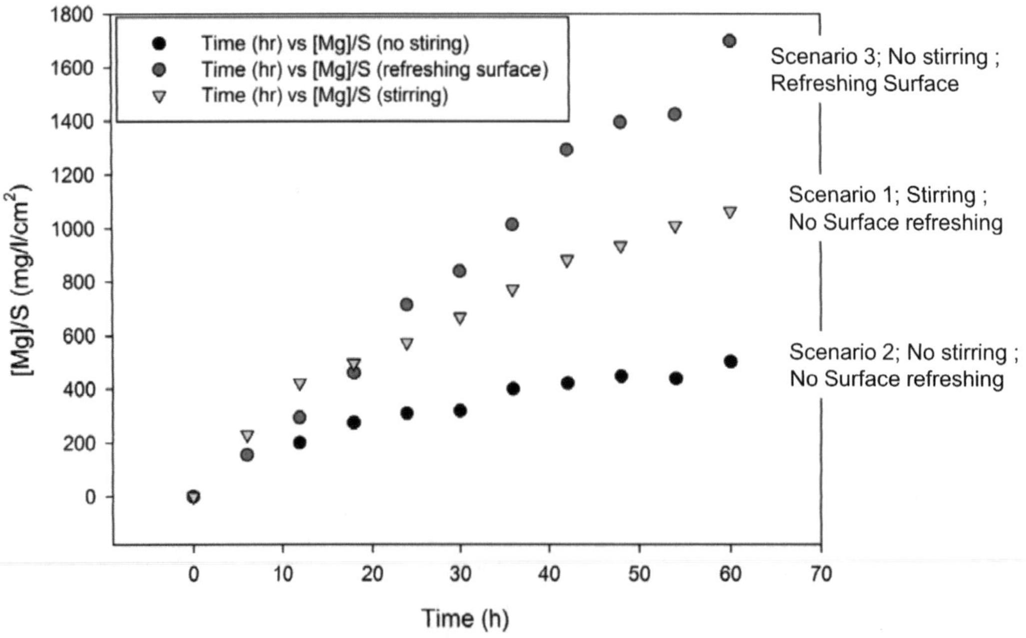

20

Scenario 1 – Shrinking Core Model (SCM)

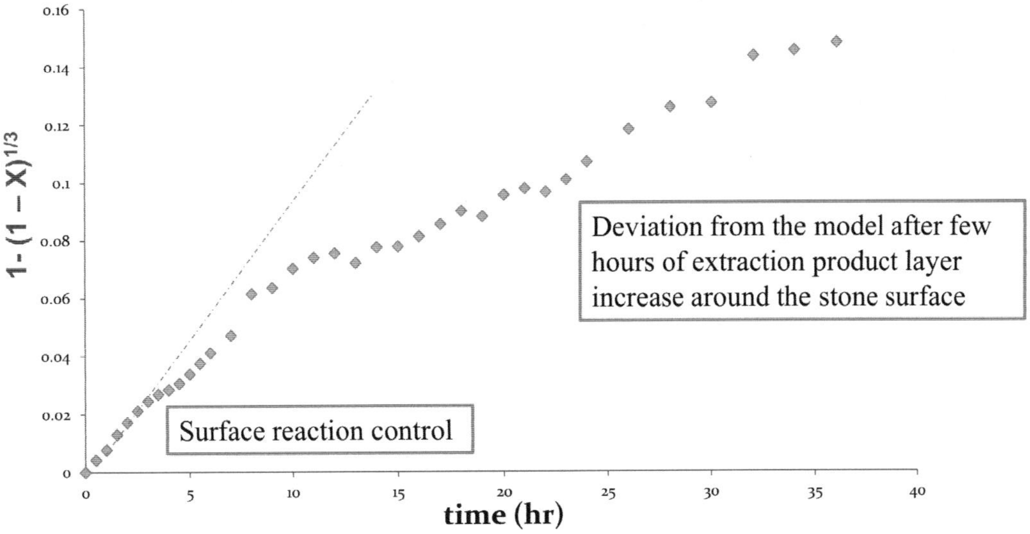

SCM Ref:

1. Yagi, S. and D. Kunii. *Studies on combustion of carbon particles in flames and fluidized beds*. in *Symposium (international) on Combustion*. 1955: Elsevier.
2. Levenspiel, O., *Chemical reaction engineering*. Vol. 2. 1972: Wiley New York

21

Fine Silica Particles Separated from the Surface of Large Particle

22

The Mechanism of Mineral Acid Dissolution
Crackling Core Model (CCM)

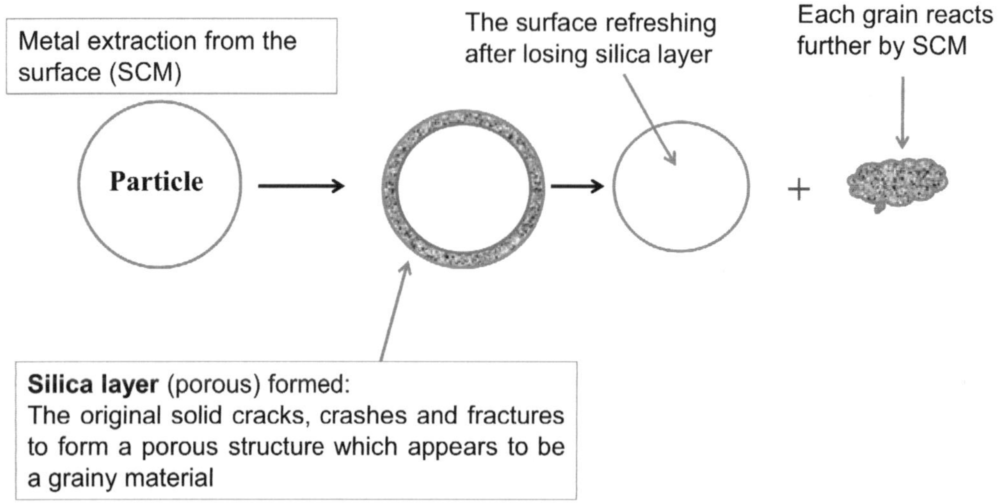

$$HCL(aq) + \frac{1}{6}Mg_3Si_2O_5(OH)_4(s) \rightarrow \frac{1}{2}Mg^{2+}(aq) + Cl^-(aq) + \frac{1}{3}SiO_2(s) + \frac{5}{6}H_2O(l)$$

23

Conclusions

- The grinding energy increases according to a power law as the mineral particle sizes get smaller. It is observed that the minerals particle could be grinded to ~500 μm with high CO2 sequestration efficiency and low grinding energy penalty.
- The dissolution mechanisms follow the crackling core model for acid-leaching process. Chemical reaction control is important at the start of the dissolution, but the silica layer control dominates at the end.
- Acid-leaching processes do not require very small particle sizes, since the crackling core mechanism eventually results smaller particles and higher conversions.

24

Any question?

411061 Investigation of Carbon Content Analysis of Chlorella vulgaris and Nannochlropsis Oculata Using TOC Analyzer for CO2 Bioremediation and Biofuel Production

Abstract:

Global warming is become a current issue that world seek to find a solution to it through mitigation of the greenhouse gases. Carbon dioxide (CO_2) is one of the well-known greenhouse gases, which mainly came from human activities because of industrialization. Different CO_2 capture strategies have been suggested and the most attractive one is the combined carbon capture and wastewater treatment using microalgae. Microalgae can utilize CO_2 to increase its biomass which can be converted to a valuable biofuel products by conversion methods whether its thermochemical, biochemical, or chemical reaction based technology. Thus, characterization of the microalgae biomass for its carbon content is a decisive criterion for the selection suitable of conversion method. TOC analyzer is used to measure total organic carbon for *Chlorella vulgaris, Nannochloropsis oculta,* microalgae species. The effect of CO_2 concentration in the growth of microalgae and the amount of carbon produced is investigated. The increasing of the carbon content during microalgal biomass growth is well determined.
Acknowledgements: The authors would like to acknowledge the support provided by King Abdulaziz City for Science and Technology (KACST) for funding this work through project No. # KACST-ARP- A-T-23-26 (KACST-ARP#T-K-11-0431).

424086 Effect of B-Site Cu-Doping on $La_{0.75}Sr_{0.25}FeO_3$ Perovskites for the Conversion of Carbon Dioxide to Carbon Monoxide

Abstract:

The effect of Cu- doping on the conversion of CO_2 to CO was investigated through the Reverse Water Gas Shift - Chemical Looping process on $La_{0.75}Sr_{0.25}FeO_3$. Six perovskite powders were synthesized $La_{0.75}Sr_{0.25}Fe_{1-Y}Cu_YO_3$ (Y = 0, 0.1, 0.25, 0.5, 0.75 and 1) and characterized through X-ray diffraction, temperature-programmed oxygen vacancy formation, and temperature-programmed reduction (TPR). Incorporation of Cu facilitates the formation of oxygen vacancies at lower temperatures but also increases the instability of the perovskite. On the Y= 0.25, 0.1 and 0 samples, temperature programmed CO_2 conversion (TPO-CO_2) after isothermal H_2-reduction at 450 °C and post-reduction XRD were performed to evaluate the ability of the materials to convert CO_2 at the lowest temperature and to identify the crystalline phases active in the reaction. Conversion of CO_2 to CO was achieved 30 °C lower on the Y=0.1 sample versus the Y=0, but less CO was produced. An analysis of potential mass transfer limitations inferred that the reaction is not mass transport limited.

Experimental methods show that, at low Cu-dopings, the formation of oxygen vacancies under inerts exhibits a contrary trend to oxygen vacancies formed under H_2. DFT calculations on the Y=0, 0.1 and 0.25 samples suggested favorable generation of low amounts of oxygen vacancies by the Y=0.1 sample, and favored generation of high amounts of oxygen vacancies by the Y=0.25 sample

426761 CO2 Capture Using Highly Porous N-Doped Carbons Impregnated with Sodium

Abstract:

This talk presents the selective CO_2 uptake capacity of sodium-impregnated and N-doped carbon sorbents (SNSs). The SNSs were synthesized by a two-step process of KOH activation of polyacrylonitrile (PAN) and subsequent NaOH impregnation of the activated N-doped carbons. Especially, SNS2-20 synthesized with the ratio of KOH to PAN equal to 2 at 700 °C and 20 wt% NaOH based on the mass of PAN exhibited a maximum CO_2 uptake of 6.84 and 4.48 mmol g^{-1} at ambient pressure, and an exceptional CO_2 capacity of 3.03 and 1.90 mmol g^{-1} under flue gas conditions (0.15 bar CO_2) at 0 and 25 °C, respectively. These enhanced CO_2 capture performance is mainly due to the existence of narrow micropores (< 0.7 nm) and intrinsic N contents together with the strong basic sites generated by NaOH impregnation. In addition, it showed CO_2/N_2 selectivities of 59.5, 68.9 and 79.4 at 0, 25 and 50 °C for the gas mixture (CO_2:N_2=15:85). We will demonstrate the unaltered CO_2 uptake subject to even consecutive adsorption–desorption cycle experiments at 25 °C. Considering its simple preparation and superior CO_2 capture properties, SNS2-20 might be a promising sorbent for practical applications of CO_2 capture and storage.

432245 Recent Advancements in Metal-Organic Frameworks (MOFs) for CO2 Capture Applications

Abstract:

Recent advancements in metal-organic frameworks (MOFs) for CO_2 capture applications

Muhammad Ilyas*,

*Center for Environment and Water Research Institute, King Fahd University of Petroleum and Minerals, Dhahran, Saudi Arabia

ABSTRACT
As CO_2 is one of the major constituent of greenhouse gases responsible for global warming which is the current major environmental concern of the world. Major source responsible for CO_2 emission is power plants burning fossil fuel in the form of coal, oil and natural gas. Nanotechnology is one of the appropriate methods being explored worldwide for CO_2 capture process. Nano-materials are being developed to trap CO_2 molecules. Metal-organic frameworks (MOFs) have emerged as promising field for efficient and economical solution towards CO_2 capture. Metal-organic frameworks (MOFs) have been identified by US Department of Energy as most effective future technology for CO_2 capture. This research reviews the details of metal-organic frameworks (MOFs), their selectivity and properties which make them applicable for CO_2 capture applications.

Keywords: MOFs, CO_2 capture, nanomaterial's

435432 Life in Plastics

Abstract:

If you take the human existence as a black box. Then many inputs that go into this box and then there is also many outputs that come out of the box. There is also accumulation and growth within the box. The one thing that is clear is there is a large consumption of natural resources and energy that goes into this black box and there is also a large waste stream which is an output. If we take a step back and see how this human operation work, we can try to re-engineer the process to find more efficiencies.
I will start by telling the story of the history of waste.
A few of the earliest known occurrences of deal with waste can be found as early as 3000 B.C. Around this time first landfill was developed when Knossos, Crete digs large holes for refuse where garbage was dumped and filled with dirt at various levels. In 2000 B.C. China developed methods of Composting and Recycling, and recycling bronze for later use. And in 500 B.C. in Athens, Greece a new law claiming garbage must be dumped at least one mile from the city was implemented.

Now we are in 2014 and much has changed with the new innovations in technology, communication and infrastructure. We now have the necessary tools to re-think waste the old ways of doing things.
My current focus is on plastic waste, both industrial and consumer. Plastics have an interesting life-cycle that spans from fossil fuels from which they are made to landfills which they end up. If it is possible to recycle and reuse these resins in an energy efficient way then this is the best way to continue their life-cycle, however due to the mixed nature of plastics this is often difficult as there are many different types of resins as well as colors. The solution that I propose is to convert this mix waste stream into usable hydrocarbons which can be used as fuels or chemical feed-stocks.

In Salt Lake City, Utah we are doing just that. We have spent the last 4 years in collaboration with the department of chemical engineering at university of Utah developing a commercial process which converts mixed plastic waste streams into high quality low sulfur fuels. I will be presenting this journey with you.

423426 Investigation of Woody Biomass and Solid Residue Generated from Anaerobic Digestion and Their Allocation to Potential Energy Generation

Abstract:

In this work, a two-stage municipal solid wastes (MSWs) disposal process has been proposed and its feasibility study has been carried out. At first stage, anaerobic digestion of municipal solid waste, such as grass, leaves and animal manure, was conducted by mixing with anaerobic sludge. Cumulative biogas yield of grass and chicken manure through high-solid co-digestion was 373 mL/g volatile solid (VS) with initial total solid (TS) 20wt% at the temperature of 35°C within 40 days. After digestion, the solid residue was dried until the moisture decreased to 40 wt%. TG analysis results showed that 86wt% of dried residues were volatile solids which could be used as feedstock for gasification. At second stage, the dried solid residue was co-gasified with woody biomass for syngas generation in a fixed-bed downdraft gasifier. Co-gasification of woody biomass and sewage sludge has already been done in our preliminary study. The experimental results showed that 20 wt. % dried sewage sludge in the feedstock was effectively gasified to generate producer gas comprising over 30 vol. % of syngas with an average lower heating value of 4.5 MJ/Nm^3. Parameter studies were performed on the basis of optimizing syngas generation of this two-stage process. These included the effects of mixing ratio, heavy metal content and moisture content on the gasification performance. This two-stage process has also been compared with single one stage of anaerobic digestion and gasification process in terms of energy conversing efficiency. Through the two stage process, full utilization of MSWs can be realized to produce biogas and syngas through anaerobic digestion and gasification.

423466 Bio-Power Generation Characteristics of a Spark-Ignited Engine Operated on Syngas Derived from Biomass Downdraft Gasification

Abstract:

The power production from renewable sources must increase intensely to meet the growing demand of power across the globe. This study focusses on evaluating the operational performance of a 10 kW Spark Ignited (SI) engine operated on syngas produced from downdraft gasification of biomass. A commercial natural gas based SI engine is modified to accommodate biomass syngas and for the measurement and control of air-fuel ratio. The load variant from 1 to 5 kW was regulated by an electric load bank to evaluate the engine operational characteristics. The continuous operation of the gasifier and syngas cleaning system were monitored to analyze quality of the syngas that was injected into the engine. The variation of the biomass feedstock was observed to investigate its impact on syngas quality and engine operation. The major combustible elements of the syngas such as CO, H_2, and CH_4 were analyzed. Output power, engine efficiency, and exhaust emissions such as CO_2, CO, NOx, SO_2, H_2S, and hydrocarbons were also examined. The outcome of this study provides the engine performance operating with 100% syngas prior to the commercial scale-up for OSU's large bio-power generation system

436250 Recovery of Energy-Dense Hydrochar and Sugars from Sweet Sorghum Bagasse

Abstract:

Sweet sorghum, a C4 crop in the grass family has potential to produce ethanol biofuel due to its high carbohydrate content and high biomass productivity. It consists of 8% seed head, 19% leaf matter, 36% bagasse, and 37% juice and can be cultivated on marginal lands, has low input requirements, and is adapted to nearly all temperate climates. Sweet sorghum accumulates fermentable sugars in the stem which is extracted in a conventional roller press and fermented into ethanol using yeast and engineered microbes. A large amount of biomass residue (or bagasse) is left behind the extraction unit operation and its utilization will be critical to sustainability of ethanol production from sweet sorghum. Hydrothermal carbonization (HTC) is an effective technology for converting waste biomass and crop residues into a uniform, energy-dense solid fuel that has improved storage/handling characteristics. This study evaluates the HTC of sweet sorghum bagasse into an energy-dense hydrochar for subsequent energy use as fuel or further conversion by combustion/pyrolysis and a sugar rich water soluble product. HTC treatment of three varieties of sweet sorghum bagasse, namely, Theis, M81-E, and Dale, was performed in pressurized batch reactors at 175-275oC and 30 min residence time. HTC products were separated into hydrochar, water soluble phase, and gases, which were analyzed by standard laboratory methods. HTC resulted in 45-65% hydrochar that had 15-20% higher energy than the raw bagasse. The water soluble products showed about ~10,000 ug/mL sugars that consisted of glucose, sucrose, and fructose. Pelletization of hydrochar generated robust pellets that showed superior qualities than pellets from woody biomass feedstocks.

410260 Biofuel from Spent Coffee Grounds

Abstract:

As the fossil fuels reserves around the world are being depleted, the need for an alternative renewable source of fuel is evident. Currently, biodiesel serves as a promising source of renewable fuel. However, it is a fairly expensive process with its feedstock comprising the major cost. Therefore, the search for alternative feedstocks is important and spent coffee grounds (SCG) can be a large and economic source of vegetable oil. The present study involves the optimization of coffee oil extraction with hexane solvent and subsequent transesterification to biodiesel. This reaction was studied using inorganic base materials as catalysts, as well as the enzyme Lipase, both in solution and immobilized. The results so far indicate that an oil yield of 20 weight % from dry SCG can be obtained, and conversions to biodiesel of up to 50% can be achieved with both forms of the enzyme catalyst, with methanol or ethanol as reactants. The immobilized lipase is prefered because it can be used for extended times in a packed bed reactor (PBR) without production of undesired soap. The reaction kinetics were also studied in a lab scale PBR.

427591 Enhanced Electrochemical Performance of Waste Tyre Recycled Activated Carbon As an Anode in Li-Ion Battery

Abstract:

With the ever increasing global demand for energy supply, it has become imperative to address the issue of energy and environment sustainability. In this work, we present an effective synthetic procedure to obtain high purity activated carbon from pyrolitic tyre char and demonstrate its performance as a high capacity anode in Li-ion battery. Carbon obtained after pyrolysis of waste tire crumbs is first treated with concentrated HCl, followed by HF treatment which converts it to a high purity carbon black devoid of all metallic and sulphur impurities. The surface properties of the acid treated carbon is further tailored by activating it with KOH, resulting in a high specific surface area of 870 m^2/g with a high percentage of mesopores (avg pore size 3-5 nm). The morphology, structure and composition of the activated carbon have been studied by SEM, TEM, Raman, XRD, XPS and FT-IR techniques. Its electrochemical performance is analyzed by galvanostatic charge/discharge measurements at different current densities, cyclic voltammetry and impedance measurements. The activated carbon exhibits a first diacharge capacity of 880 mAh/g at 50 mA/ g current density, which is much greater than other (hard) carbon based anodes and more than twice that of commercial graphite anode. A high reversible capacity of ~ 620 mAh/g is maintained after 100 cycles with nearly 100% coloumbic efficiency. Our work opens a new avenue to tyre waste recycling as the carbon black processed by us exhibits enhanced electrochemical performance with potential application in energy storage devices.

433364 Catalytic Conversion of Automobile Shredder Residue to High Energy Fuels

Abstract:

Catalytic pyrolysis of automobile shredder residue to high energy fuels

FA Agblevor, K. Christian, S. Hatch, J. Long

USTAR Bioenergy Center, Biological Engineering, Utah State University, Logan UT 84322

Abstract

Automobile shredder residue (ASR) or auto-fluff is waste product from the extraction of metals from automobiles. The material is composed of plastics, vinyl, cellulose, wood, metals, leather, and other materials. ASR is currently disposed in public landfills. However because of the organic material content of this residue, it is potentially possible to convert the material to fuels using pyrolysis technology. ASR and red mud samples were characterized using several methods. The ASR samples showed large variability and had very high ash content. The ASR samples were pyrolyzed using sand and red mud. Pyrolysis oils obtained using sand as the pyrolysis medium were more viscous than those using red mud and they also contained polychlorinated biphenyls (PCB) at high levels. The yield of ASR pyrolysis liquids ranged from 2 to 32 wt%. Pyrolysis with red mud produced much lower viscosity oils and which also had very low PCB content. Whereas there were several types of PCBs in the sand pyrolysis oils, there was only one type PCB congener (trichlorobiphenyl) in the red mud ASR pyrolysis oils. Modification of the red mud with reduced the PCB content of the ASR red mud pyrolysis oils to 5 ppm compared to over 50 ppm in the sand ASR pyrolysis oils. The red mud ASR pyrolysis oils had close to neutral pH while the sand pyrolysis oils were acidic. The gas yield ranged from28 to 36 wt%. The red ASR pyrolysis gas was H2-rich while the sand catalyzed pyrolysis oil was CO2-rich. The char-yield was very high and ranged from 38 to 42 wt% and had ash content over 70%. The red mud ASR pyrolysis oil in its current state can be potentially used as heating fuel but needs upgrading to qualify as transportation fuel.

430405 Hydrothermal Carbonization for Nutrient Recovery

Abstract:

A variety of macro- and micronutrients are required for plant's healthy growth and maximum crop production. The most common macronutrients associated with commercial fertilizers are nitrogen, phosphorous, and potassium (NPK), each of which is commercially available in large quantities for modern farming. Until now, NPK containing fertilizers are produced by either accompanying chemical processes (e.g., Haber-Bosch process) or physical processing of mineral ores (e.g. potash). Although, most of the common waste streams (e.g. sewage sludge, dairy manure etc.) are very rich in macro-and micronutrients, the lack of economically and environmentally benign processes is hindering the use of nutrients from wastes.

Hydrothermal carbonization (HTC or wet torrefaction) is a pretreatment process, which makes a homogenized, carbon and nutrient rich solid product, often called hydrochar, from underutilized various wet feedstocks, including dairy manure and sewage sludge. Compared to raw biomass, hydrochar is more hydrophobic and more biologically inert, which favors better storage properties for further applications. In this pretreatment method, biomass is treated with hot compressed water in an inert environment in the temperature range of 180-260 °C. The ionic constant of water is highest in this temperature range. As a result, the oxygen containing volatiles are removing from the solid, leaving a stable carbon and nutrient rich solid hydrochar.

In this study, macronutrients especially NPK recoveries from various wastes were evaluated by HTC. Both dairy manure and sewage sludge were hydrothermally carbonized at 180-260 °C for 5-30 min. Physico-chemical properties of both solid and liquid products were determined by means of ICP-OES, Lachat, TOC/TN, and elemental CHONS analyses. Total nitrogen and NH_4-N in the HTC process liquid was increasing with HTC temperature and time. However, more than half of the starting nitrogen was recovered in the hydrochar. Phosphorus recovery in the solid hydrochar was also increasing with HTC severity but potassium was found entirely in the process liquid. In conclusion, HTC can be applied to extent for nutrient recovery especially for nitrogen and phosphorus in the solid product.

435161 Production of Bio-Diesel Additives from Dairy Manure in Supercritical Ethanol

Abstract:

Production of Bio-diesel Additives from Dairy Manure in Supercritical Ethanol

Xiaokun Yang[1], Mi Lu[1], Marcus Vinicius Pereira[1], Simon Poulson[3], Charles Coronella[1], and Hongfei Lin[1*]

[1]Chemical Engineering and Materials Science Department, [2]Agriculture, Biotechnology & Natural Resources Department, [3]Geological Sciences & Engineering Department,

University of Nevada, Reno, NV, U.S 89557

E-Mail: HongfeiL@unr.edu

Dairy manure is regarded as a waste material that needs substantial waste management. However, as the biomass feedstock it contains lipid, protein, lignin, hemicellulose, and cellulose, which can be used to produce value-added chemicals. Herein, dried dairy manure was converted in ethanol and ethanol-water solutions at near supercritical conditions. The highest conversion of manure reached to 72% (ash free) under supercritical ethanol at 245°C. In the presence of solid Lewis acid catalysts, Zr-KIT-6 or hydrated niobia, ethyl esters from cellulose, hemicellulose and lignin were the main products, which can be used as biodiesel additives. It was found that adding 5-10 % water in ethanol enhanced the decomposition of recalcitrant cellulose and lignin, and thus increased the manure conversion and the overall yields of ethyl esters. Various process conditions including temperature, pressure, biomass-to-catalyst ratio, water-to-ethanol ratio, etc. were examined and optimized. High-loading amounts of manure were used in this process which makes the catalytic process promising for future's scale-up to industrial scale. Lastly, the possible reaction mechanism based on Lewis acid catalysis in supercritical ethanol-water solvents was discussed.

423642 Carbothermal Zinc Recycling in a Packed-Bed Solar Reactor – Experimental Demonstration and Dynamic Modeling

Abstract:

Carbothermal Zinc Recycling in a Packed-Bed Solar Reactor – Experimental Demonstration and Dynamic Modeling

Waelz oxide (WOX) is produced in the dominating Zn recycling pyrometallurgical process, the so called Waelz process. The crude WOX generated contains 55–65% Zn in oxidic form, along with considerable amounts of chlorine, fluorine and lead compounds. A washing treatment for the production of dehalogenized WOX (washed WOX) is thus required before using WOX as a feedstock along with ZnS concentrates for the electrolytic production of primary Zn.

In the present study, an environmentally cleaner process route for the purification of crude WOX is experimentally investigated using concentrated solar process heat [1]. The solar-driven process comprises the thermal purification of WOX via volatilization of the Cl-, F- and Pb-compounds at temperatures between 1000 and 1285 °C to yield solar-clinkered WOX, which contains approximately 80 wt.% ZnO and can be used as feedstock for the subsequent carbothermal reduction.

Experimentation was carried out using a 10 kW_{th} packed-bed solar reactor [2]. It consists of two cavities in series separated by a thin ceramic plate, with the upper cavity serving as the solar radiative absorber and the lower cavity containing the reacting packed bed of solid feedstock. The laboratory scale reactor was subjected to solar flux concentrations of up to 2,500 suns (1 sun = 1 kW/m^2). Solar-driven purification of WOX at above 1265°C reduced the amount of Cl- and Pb-compounds to levels lower than those of washed WOX (< 0.1 wt.%) and led to ZnO contents of approximately 88 wt.% in the product.

The second process investigated was the solar-driven carbothermal reduction of crude, washed, and solar-clinkered WOX as well as commercial ZnO powder. Beech charcoal was used as reducing agent for a CO_2-neutral carbothermal process. Solar experimental runs were carried out with the 10 kW_{th} packed bed reactor at temperatures up to 1170 1320°C and ZnO:C molar ratios between 1:0.8 and 1:1.25. Condensed products contained 90 wt.% Zn.

A dynamic reactor model was developed to solve numerically the transient energy conservation equation coupling heat transfer with the reaction kinetics. Model validation was accomplished by comparison of packed-bed temperatures and Zn production rates with experimentally measured data obtained with the 10 kW_{th}packed-bed solar reactor. The validated model was further applied for optimizing the solar reactor configuration.

References

1. Wieckert C, Frommherz U., Kraeupl S., Guillot E., Olalde G., Epstein M., Santen S., Osinga T., Steinfeld A., "A 300 kW solar chemical pilot plant for the carbothermic production of zinc", *ASME Journal of Solar Energy Engineering*, Vol. 129, pp. 190-196, 2007.
2. Tzouganatos N., Matter R., Wieckert C., Antrekowitsch J., Gamroth M., Steinfeld A., "Thermal Recycling of Waelz Oxide using Concentrated Solar Energy", *Journal of Metals*, Vol. 65, pp. 1733-1743, 2013.

429396 Greenscope Tool for Process Sustainability Assessment and Life Cycle Inventory Generation

Abstract:

The development of sustainable solutions for addressing the satisfaction of the human population, its material and energy needs, and simultaneously preserving the Earth systems that make human life possible has been an objective of researchers working in academia, government, and industry for decades. Commonly, new chemicals and their manufacturing processes are developed, however, some evaluation and analyses are needed for supporting the classification of these new contributions as sustainable alternatives. This contribution demonstrates the application of the U.S. Environmental Protection Agency's software tool, GREENSCOPE, for the assessment of process sustainability and life cycle inventory generation. GREENSCOPE is a web-based tool for evaluating sustainability of existing and new technologies by employing performance indicators for objectives in four main areas: environment, economics, energy, and efficiency. A case study demonstration using GREENSCOPE will be performed to provide users guidance on using this assessment and design tool, from data entries to sustainability evaluation outcomes and the related life cycle inventory generation. The indicator scores describe product and process performance in a sustainability measurement scale, determining whether chemical products and processes are more or less sustainable.

412272 LCA Needs to Move Toward Sustainability to Remain Relevant-

Abstract:

LCA, a methodology amply utilized to environmentally assess product systems under the requirements of ISO LCA standards, has been used for the last decade or so with important constraints or limitations. Among the most important of such limitations are specifically its use in comparing product systems and in its use as a predictor of sustainability. Albeit the issues around its capacity to accurately and fairly compare product systems have been amply discussed and the original standard revised or modified in those sections having to do with comparative assertions, etc. still. limitations as a predictive tool for environmental preference among product systems persist. However, the LCA standard capability to address sustainability issues has not been examined in the same depth and extent. Nevertheless, the demand of society for a proper indicator of sustainability of different product systems has resulted in the improper suggestion, and even use, of inadequate methodologies. LCA, because its popularity, has been considered one such a metric. Another, purely arithmetic reduction efficiency, has and its being used as a pseudo sustainability metric. In both cited approaches, the lack of context and thresholds invalidate its valid use as sustainability metrics. The paper analyzes the fundamental characteristics of LCA methodology in ISO standards vis a vis the needed characteristics for a proper sustainability metric for product systems. The results of the analysis clearly indicate that LCA study results are not proper metrics to evaluate sustainability of products systems. A more proper and effective use of LCA would be as a preliminary analysis of the product system from which to focus in more important elements or indicators of the LCA study analysis. A concerted effort should be initiated by academy and practice to incorporate needed thresholds or pseudo-thresholds, in the methodology and practice of LCA. As now, in general, the results of a LCA study, by themselves, are improper metrics for sustainability.

436855 Sustainability Metrics for Charging Infrastructure for Electric Vehicles

Abstract:

Sustainability metrics have value in life cycle analysis and the evaluation of alternatives. One approach to expanding the infrastructure for electric vehicles (EVs) is to install solar powered charging stations with solar panels above parking spaces to generate electricity and provide shade for cars parked under the panels. Grid connected systems with charging stations enable electricity to flow into the grid or into the batteries of the EVs. Social, environmental, and economic metrics are included in the analysis of benefits to society of an infrastructure of solar powered charge stations. Social metrics include the social value of shade, convenience, and availability, and better air quality. Environmental metrics include greenhouse gas emissions, air quality, water use, and land requirements. Economic metrics include capital costs, EV charging cost, and cost of transportation. Several alternatives will be compared.

NRel
NATIONAL RENEWABLE ENERGY LABORATORY

Combined Sustainability Assessment and Techno-Economic Analysis for the Production of Biomass-Derived High-Octane Gasoline Blendstock

Eric C.D. Tan

National Renewable Energy Laboratory (NREL) Golden, CO

November 12, 2015

Session: Sustainability Metrics at the Process and Product Level

2015 AIChE Annual Meeting, Salt Lake City, UT

NREL is a national laboratory of the U.S. Department of Energy, Office of Energy Efficiency and Renewable Energy, operated by the Alliance for Sustainable Energy, LLC.

Publications

http://www.nrel.gov/docs/fy15osti/62402.pdf

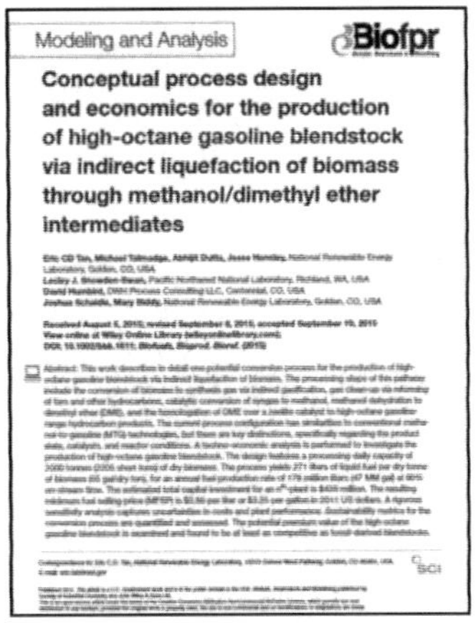

http://onlinelibrary.wiley.com/doi/10.1002/bbb.1611/abstract

http://www.energy.gov/sites/prod/files/2015/03/f20/mypp_beto_march2015.pdf

Highlights of this IDL conversion pathway

- Produces a high-octane, gasoline-range (C5+), hydrocarbon product.
- Leverages advances in syngas production and clean-up technologies developed for the thermochemical pathway from biomass to mixed alcohols, demonstrated in 2012.
- Operates at lower severity conditions (lower operating temperature and pressure) relative to traditional methanol-to-gasoline (MTG) processes.
- Combines homologation (-CH2- addition) and conversion of recycled C4-olefins and iso-paraffins to larger products in a single reactor system and does not require a separate alkylation step.
- Possesses high selectivity to C7 (mostly 2,2,3-trimethylbutane and 2,2,3-trimethyl-but-1-ene) and C4 paraffins and olefins, which can be recycled to maximize yield of gasoline blendstock product.
- Produces low aromatic-content product and has a low coke formation rate.

High-Octane Gasoline Pathway vs. MTG

Process Attribute	High-Octane Gasoline Pathway Target	Methanol to Gasoline (MTG) Pathway	Impact on Techno-Economic Analysis
Molecular structures favored in synthesis reactions Example Compound Specific Gravity	Branched paraffins 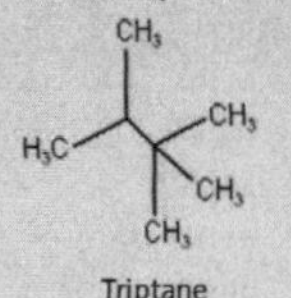 Triptane 0.70	Aromatics 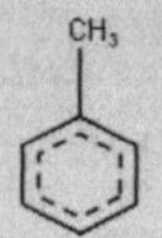 Toluene 0.87	High octane product rich in branched paraffins, similar to a refinery alkylate. H-saturation decreases density, increasing product volume.
Hydrocarbon synthesis catalyst	Beta-Zeolite (12-membered rings)	ZSM-5 (10-membered rings)	Different pore sizes and structures result in different compound selectivities.
Octane number of gasoline-range product	RON: 95+ MON: 90+	RON: 92 MON: 83	Octane number increases value of product as a finished fuel blendstock.
Selectivity of C_5+product	C_5+ product only (~65 Gal / Ton)	~ 85% C_5+ (~55 Gal / Ton)	High selectivity to primary (premium quality) product maximizes overall product value.
Severity of synthesis operating conditions	350 – 450 Deg. F 130 PSIA	650 – 950 Deg. F 315 PSIA	The lower severity operating conditions result in lower capital and operating costs relative to MTG.
Coke formation	Coke formation is minimized by hydrogen addition and selectivity to branched paraffins rather than aromatics.	High propensity for coke formation due to aromatic coke pre-cursors.	Minimizing coke formation helps to maximize product yield / carbon efficiency and maximizes catalyst regeneration and replacement cycles.

Sources: (1) 2011 NREL MTG Design Report (Phillips et al), (2) Methanol to Gasoline (MTG) Production of Clean Gasoline from Coal (ExxonMobil Research and Engineering)

Simplified Process Flow Diagram

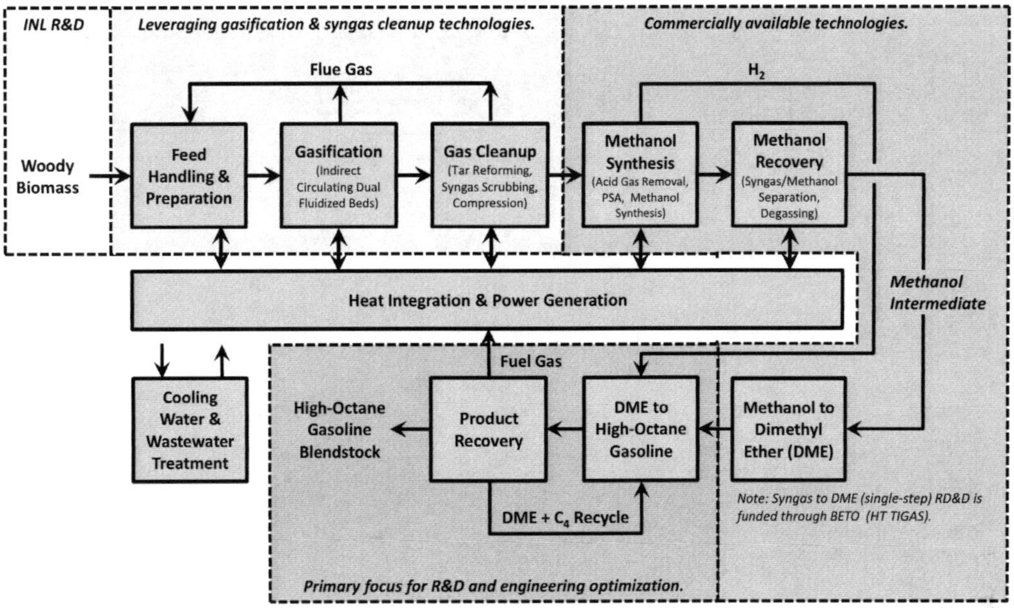

Gasification and Gas Clean-up

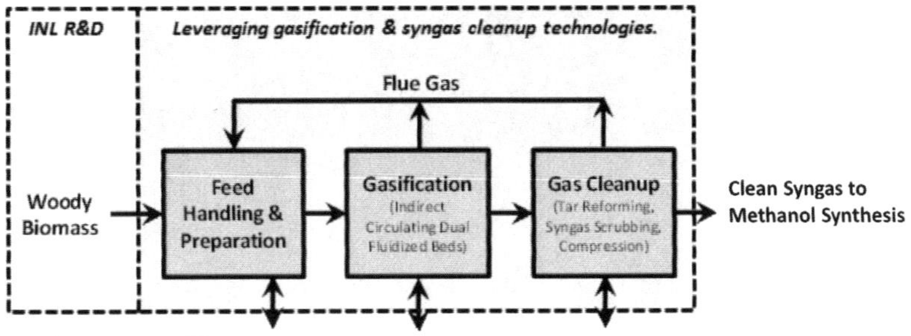

Indirect Gasifier

- INL woody feedstock blend (<1% ash, <10% moisture). Waste heat required only to preheat the feedstock
- Consistent with 2012 mixed alcohol design / demonstration

  Dutta, A., et al. "Technoeconomic Analysis for the Production of Mixed Alcohols via Indirect Gasification of Biomass Based on Demonstration Experiments," Industrial & Engineering Chemistry Research, 53(30), 12149-12159 (2014).

Gas Cleanup

- Tar reforming (based on 2012 mixed alcohol design / demonstration)
- Scrubbing to remove residual fines

Acid Gas Removal & Methanol Synthesis

Acid Gas Removal

- Combined AGR and LO-CAT systems
- Remove CO2 and H2S
- ZnO sulfur guard bed is required

Methanol Synthesis

- Syngas further conditioned to meet CO_2 (5 mol%) and H_2S (0 ppm) specifications
- Methanol conversion based on commercial technologies
- Fixed-bed shell & tube reactor containing a copper/zinc oxide/alumina catalyst
- Operating conditions: 500 F, 730 psia
- 95 wt% of the unconverted syngas is recycled back to the synthesis reactor
- Portion of syngas is sent through PSA to recover hydrogen for use downstream

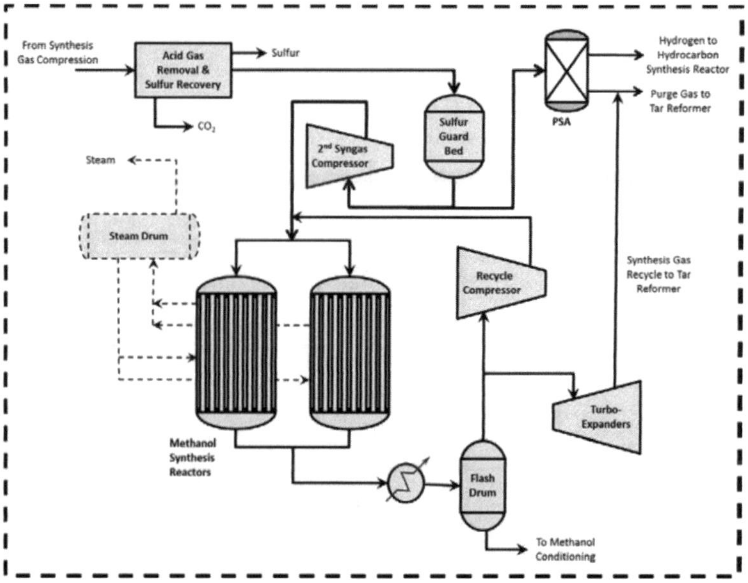

DME to High-Octane Gasoline

DME Synthesis

- Vapor-phase dehydration of methanol to DME on gamma alumina catalyst at 480 F and 140 psia.

Hydrocarbon Synthesis

- DME is converted to gasoline-range hydrocarbons over 3 parallel fixed bed reactor trains (with inter-bed cooling) containing metal-modified beta-zeolite catalyst.
- Operating conditions: 438 F, 130 psia.

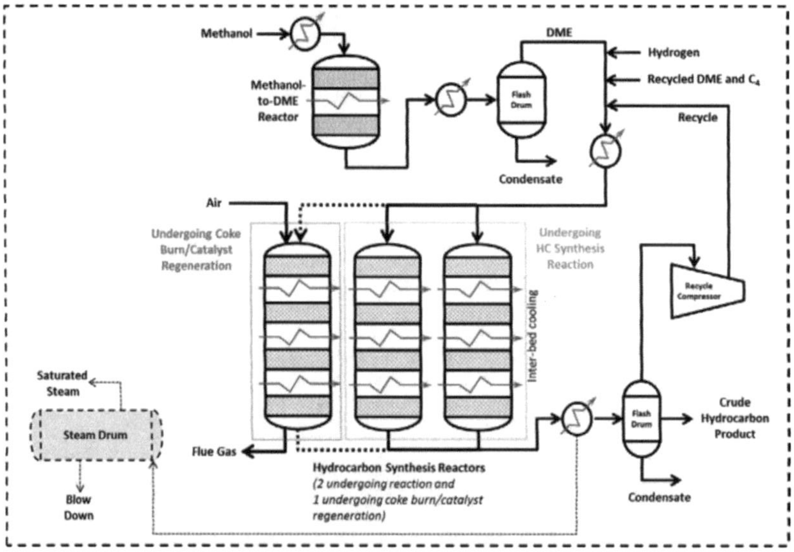

- DME conversion: 40% (single-pass); 93% (overall).
- H2 is introduced to reduce aromatics formation and C4s are recycled to improve yields.
- Potential for dimerization step to produce jet-range molecules.

Product Recovery

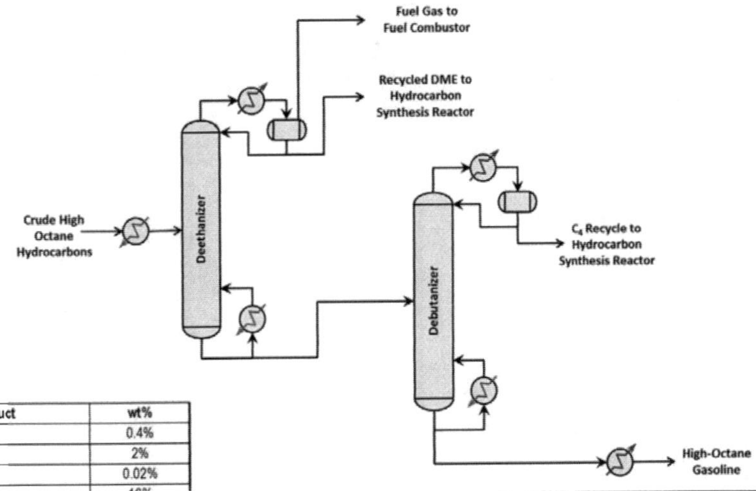

Crude Hydrocarbon Product	wt%
Light gas	0.4%
Propane	2%
Propylene	0.02%
Isobutane	16%
n-Butane	0.3%
Butenes	0.4%
C5+ gasoline	81%
Total	100%

Finished Fuel Products	wt%
Gasoline	95%
LPG	0%
Fuel gas	5%
Total	100%

Typical Crude Gasoline Separation

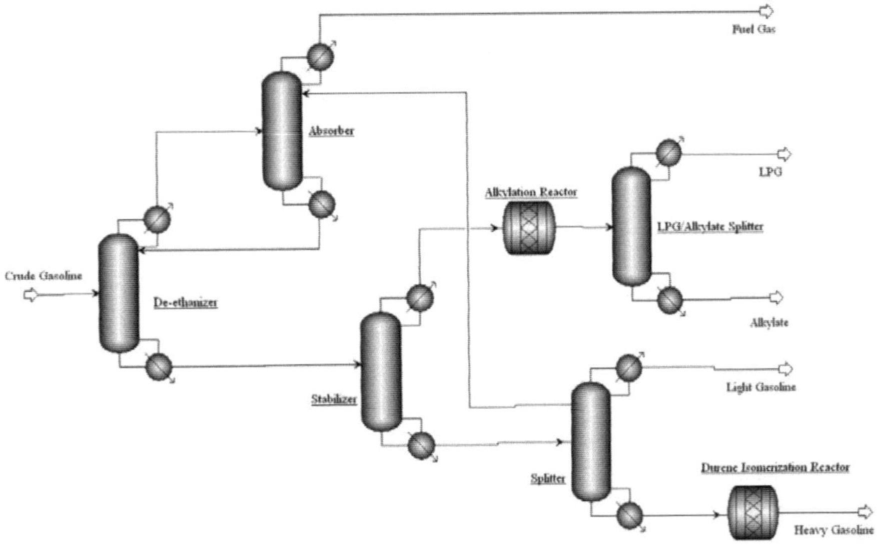

Source: Phillips, S.D.; Tarud, J.K.; Biddy, M.J.; Dutta, A. Gasoline from Wood via Integrated Gasification, Synthesis, and Methanol-to-Gasoline Technologies. NREL/TP-5100-47594. Golden, CO: National Renewable Energy Laboratory, 2011.

Comparison of gasoline composition

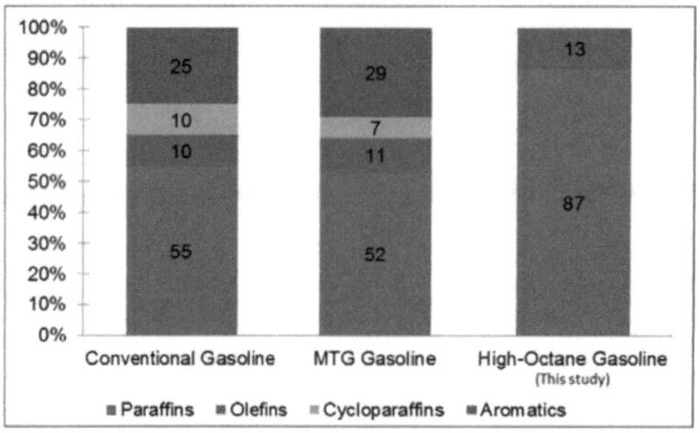

- Gasoline produced from this high-octane gasoline pathway contains only paraffins and olefins.
- High octane rating is largely attributed to its high concentration of iso-paraffins (98%).
- As opposed to MTG, no durene (1,2,4,5-tetramethylbenzene) is formed in this high-octane gasoline process.

Operating and Financing Assumptions

Description of Economic Parameter	Analysis Value / Basis
Delivered Feedstock Cost	\$80.00 / US Dry Ton
Internal Rate of Return (IRR)	10.0 %
Plant Financing by Equity / Debt	40% / 60% of Total Capital Investment
Plant Life	30 Years
Income Tax Rate	35.0 %
Interest Rate for Debt Financing	8.0 % Annually
Term for Debt Financing	10 Years
Working Capital Cost	5.0% of Fixed Capital Investment
Depreciation Schedule	Total Plant: 7-Year MACRS
Construction Period (Spending Schedule)	3 Years (8% Y1, 60% Y2, 32% Y3)
Start-Up Time	6 Months
Revenue and Costs During Start-Up	Revenue = 50% of Normal Operation Variable Costs = 75% of Normal Operation Fixed Costs = 100% of Normal Operation
On-Stream Percentage After Start-Up	90% (7,884 Operating Hours per Year)

Design Case Economic Summary

Description of Economic Parameter	Current Value
Minimum Fuel Selling Price	$3.41 / Gal GE
Minimum Fuel Selling Price	$3.25 / Gallon
Feedstock Costs	$1.23 / Gallon
Operating Costs & Credits	$0.55 / Gallon
Capital Charges & Taxes	$1.47 / Gallon
Gasoline Blendstock Production	47.0 MM Gallons / Year
Gasoline Blendstock Product Yield	64.9 Gallons / US Dry Ton
Delivered Feedstock Cost	$80.00 / US dry ton
Total Purchase Equipment Cost (TPEC)	$122.2 MM
Installation Factor / Total Installed Cost (TIC)	x 2.06 Factor = $251.2 MM
Total Fixed Capital	$415.2 MM
Working Capital / Land Costs	$20.1 MM / $1.6 MM
Total Capital Investment	$437.5 MM
Lang Factor (TCI / TPEC)	3.58
TCI Lang Factor (TCI / ISBL TPEC)	4.54

Design Report Cost Distribution Details

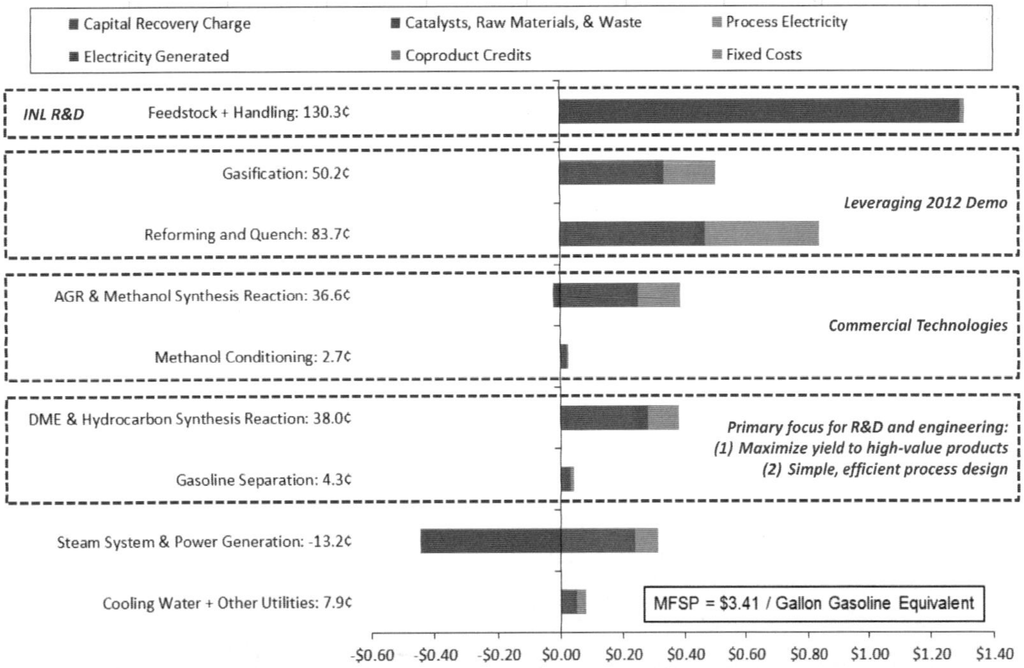

Conversion Stage Sustainability Metrics

GHGs (g CO_2-e/MJ fuel) – (fossil emissions; biogenic emissions)	0.6; 153
Fossil energy consumption (MJ fossil energy/MJ fuel product)	6.0E-03
Total fuel yield (gal/dry ton feedstock)	64.9
Biomass Carbon-to-Fuel Efficiency (C in fuel/C in biomass)	31%
Water Consumption (m^3/day; gal/GGE)[1]	911; 1.76
Wastewater Generation (m^3/day; gal/GGE)	123; 0.24

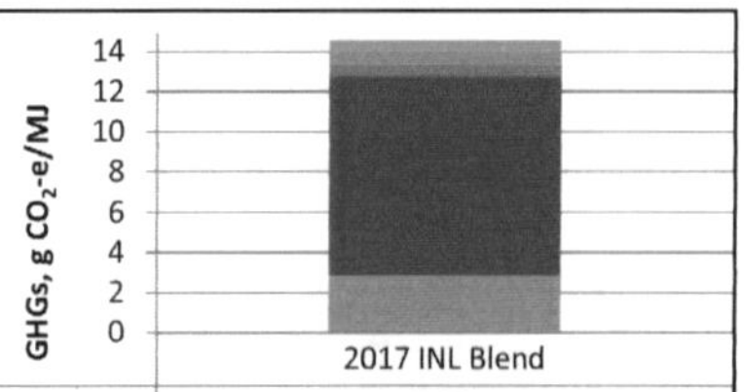

	2017 INL Blend
Fuel Use	0.49
Fuel Distribution	0.72
Conversion	0.57
Harvest, Collection and Preprocessing	10.0
Crop Establishment and Growing	2.8
Total	14.6

The overall life-cycle GHG emissions (14.6 g CO2-e/MJ from feedstock production to vehicle operation) reduction exceeds the 60% reduction criteria relative to the 2005 petroleum gasoline baseline (93.08 g CO2-e/MJ).

Single-Point Sensitivity Analyses

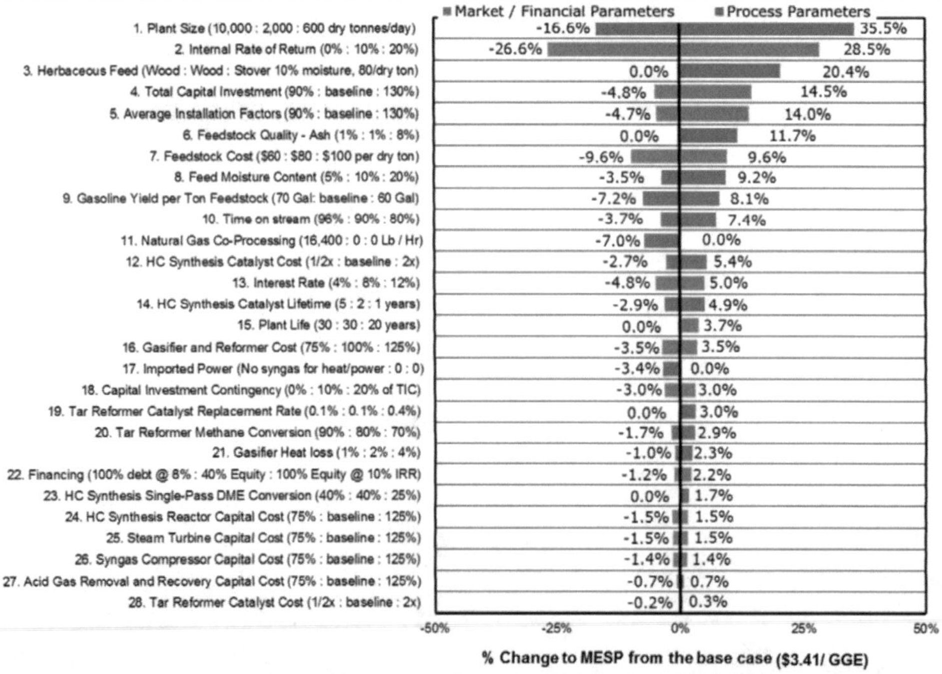

GHG/Cost Trade-Offs & Synergies: Gasification
Co-conversion with natural gas

- **Thermochemical conversion pathway to high-octane gasoline blendstock through methanol/DME intermediates**
  - Co-conversion of biomass with natural gas can simultaneously increase fuel yields and reduce fuel production costs provided that the life-cycle GHG thresholds specified in EISA is not violated.
  - Co-processing natural gas at the life-cycle GHG threshold limit (i.e., 60% GHG emissions reduction relative to the petroleum baseline) decreases the MFSP by 7%, from \$3.41/GGE to \$3.17/GGE.

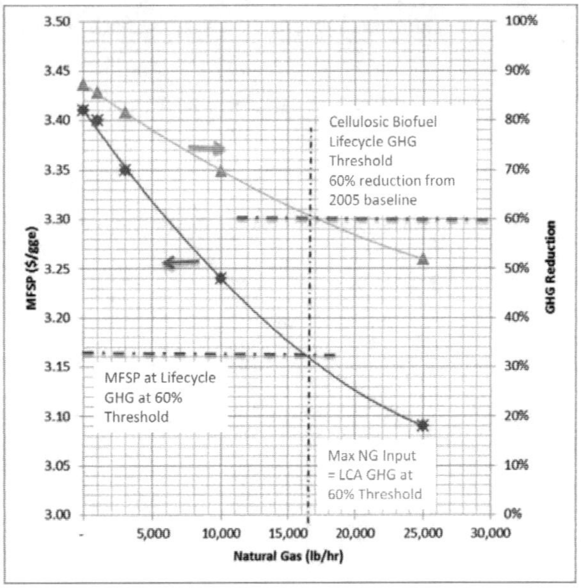

Potential Premium Value of the Design Case Modeled High-Octane Gasoline Blendstock

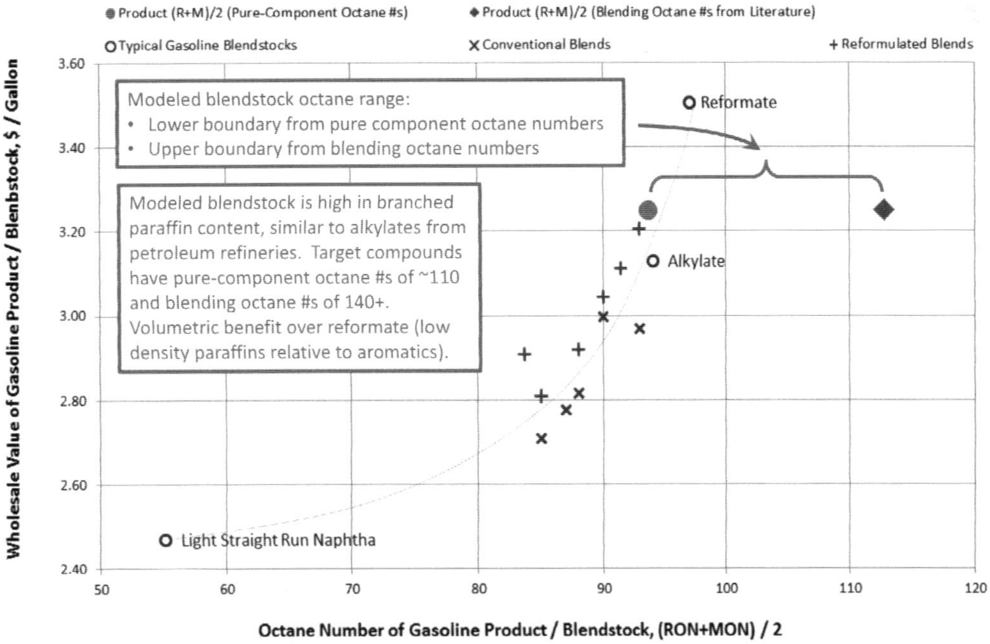

Sources: (1) May 2014 Spot Prices from EIA, (2) Platts Oilgram Price Report - July 11, 2013, (3) Modern Petroleum Technology; 5th Edition Part II; Edited by G.D Hobson, Wiley 1984.

Major Technical Targets for High-Octane Gasoline Pathway

Process Parameter	2014 SOT	2015 Target	2016 Target	2017 Target	2022 Target / Design Case
Hydrocarbon Synthesis Catalyst	Commercially available beta-zeolite		Beta-zeolite modified with copper (Cu) and gallium (Ga) for performance improvement		
H_2 Addition to HC Synthesis	No	Yes			
C_4 Co-Product (Function of Recycle)	Yes (No Recycle of C_4s)		No (Recycle C_4s)		
Single-Pass DME conversion	15%	15%	20%	30%	40%
Productivity of Hydrocarbon Synthesis Catalyst (kg/kg-cat/h)	0.02	0.03	0.04	0.05	0.10
Carbon Selectivity to C_5+ Product	46.2%	50.8%	86.1%	89.9%	93.1%
Carbon Selectivity to Aromatics (HMB represents coke / pre-cursers)	25% Aromatics (10% HMB)	15% Aromatics (7% HMB)	8% Aromatics (4% HMB)	4% Aromatics (2% HMB)	0.5% as HMB
Dimerization of C_4-C_8 Olefins to Jet	No	Consider in SOT cases as sensitivity or modify target case			

SOT / Target Waterfall for High-Octane Gasoline Pathway

Minimum Fuel Selling Price (MFSP) Breakdown ($ / Gallon of Gasoline Equivalent)	2014 State of Technology	2015 Target	2016 Target	2017 Target	2018 Target (Interpolated)	2019 Target (Interpolated)	2020 Target (Interpolated)	2021 Target (Interpolated)	2022 Target / Design Case
Feedstock	$ 1.99	$ 1.82	$ 1.48	$ 1.32	$ 1.31	$ 1.31	$ 1.31	$ 1.31	$ 1.30
Gasification	$ 0.70	$ 0.67	$ 0.56	$ 0.54	$ 0.53	$ 0.52	$ 0.52	$ 0.51	$ 0.50
Synthesis Gas Clean-up (Reforming and Quench)	$ 1.06	$ 1.00	$ 0.84	$ 0.84	$ 0.84	$ 0.84	$ 0.84	$ 0.84	$ 0.84
Acid Gas Removal, Methanol Synthesis and Methanol Conditioning	$ 0.59	$ 0.55	$ 0.44	$ 0.42	$ 0.42	$ 0.41	$ 0.41	$ 0.40	$ 0.39
Hydrocarbon Synthesis	$ 1.01	$ 1.01	$ 0.67	$ 0.60	$ 0.55	$ 0.51	$ 0.47	$ 0.42	$ 0.38
Hydrocarbon Product Separation	$ 0.05	$ 0.05	$ 0.05	$ 0.04	$ 0.04	$ 0.04	$ 0.04	$ 0.04	$ 0.04
Balance of Plant	$ 0.04	$ (0.00)	$ (0.00)	$ (0.04)	$ (0.04)	$ (0.04)	$ (0.05)	$ (0.05)	$ (0.05)
Minimum Fuel Selling Price (MFSP)	**$ 5.45**	**$ 5.09**	**$ 4.04**	**$ 3.72**	**$ 3.66**	**$ 3.59**	**$ 3.53**	**$ 3.47**	**$ 3.41**

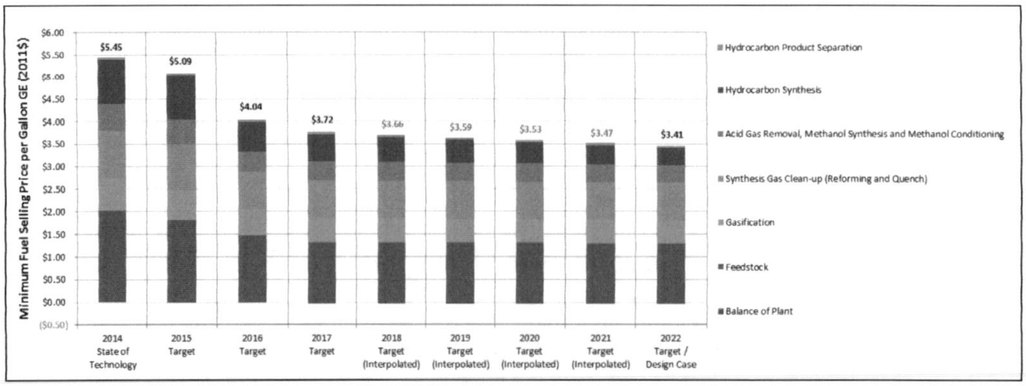

Product Quality Outlook for High-Octane Gasoline Pathway

Modeled product quality trends (1 of 3)

Product Quality Trends	2014 State of Technology	2015 Target	2016 Target	2017 Target	2018 Target (Interpolated)	2019 Target (Interpolated)	2020 Target (Interpolated)	2021 Target (Interpolated)	2022 Target / Design Case
Aromatics Content of C5+ Product (Wt%)	25.3%	15.0%	8.0%	4.0%	3.3%	2.6%	1.9%	1.2%	0.5%
C5+ Product Specific Gravity	0.75	0.73	0.71	0.70	0.70	0.70	0.70	0.70	0.70
C5+ Product API Gravity	58.17	62.41	68.57	69.37	69.52	69.68	69.83	69.98	70.14
Mixed C4 Product (Wt% Total Products)	27.2%	27.4%	0.0%	0.0%	0.0%	0.0%	0.0%	0.0%	0.0%
C5+ Aromatics (Wt% Total Products)	18.5%	10.9%	8.0%	4.0%	3.3%	2.6%	1.9%	1.2%	0.5%
C5+ Non-Aromatics (Wt% Total Products)	54.4%	61.7%	92.0%	96.0%	96.7%	97.4%	98.1%	98.8%	99.5%

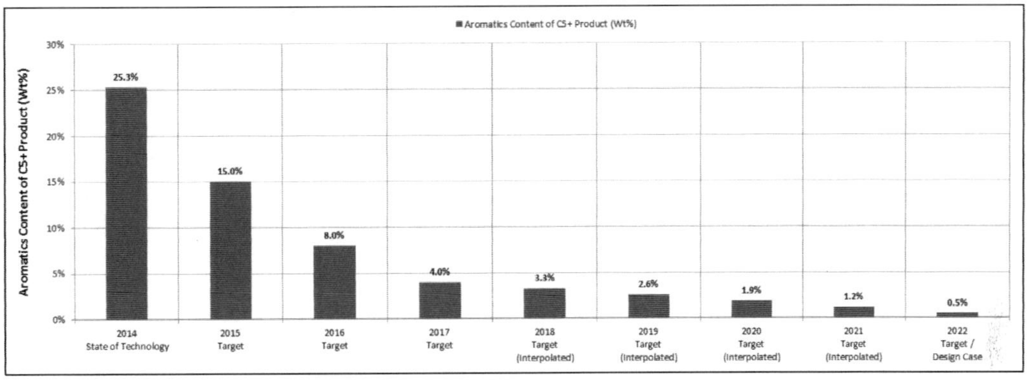

Sustainability Metrics for High-Octane Gasoline Pathway

Modeled trends in sustainability metrics (1 of 2)

Sustainability Metric Trends	2014 State of Technology	2015 Target	2016 Target	2017 Target	2018 Target (Interpolated)	2019 Target (Interpolated)	2020 Target (Interpolated)	2021 Target (Interpolated)	2022 Target / Design Case
Fossil GHG Emissions (g CO2-e / MJ Fuel)	1.64	1.42	1.19	0.96	0.88	0.81	0.74	0.67	0.60
Fossil Enegy Consumption (MJ Fossil Energy / MJ Fuel)	0.023	0.019	0.011	0.013	0.011	0.010	0.009	0.007	0.006
Total Fuel Yield (Gallons / Dry Ton)	57.5	58.5	61.8	64.2	64.4	64.5	64.6	64.8	64.9
Total Fuel Yield (Gallons GE / Dry Ton)	51.3	52.2	59.1	61.4	61.5	61.6	61.7	61.8	61.9
Overall Carbon Efficiency to Hydrocarbon Products (% C in Feedstock)	28.2%	28.7%	29.9%	31.0%	31.0%	31.0%	31.1%	31.1%	31.2%
Water Consumption (Gal H2O / Gal C5+)	12.4	9.3	5.8	5.2	4.5	3.8	3.1	2.4	1.7
Electricity Export (kWh / GGE)	0	0	0	0	0	0	0	0	0

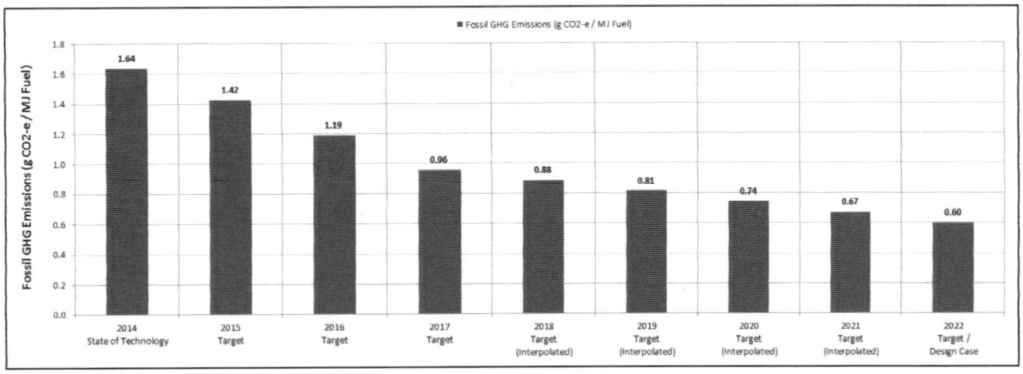

Sustainability Metrics for High-Octane Gasoline Pathway

Modeled trends in sustainability metrics (2 of 2)

Sustainability Metric Trends	2014 State of Technology	2015 Target	2016 Target	2017 Target	2018 Target (Interpolated)	2019 Target (Interpolated)	2020 Target (Interpolated)	2021 Target (Interpolated)	2022 Target / Design Case
Fossil GHG Emissions (g CO2-e / MJ Fuel)	1.64	1.42	1.19	0.96	0.88	0.81	0.74	0.67	0.60
Fossil Enegy Consumption (MJ Fossil Energy / MJ Fuel)	0.023	0.019	0.011	0.013	0.011	0.010	0.009	0.007	0.006
Total Fuel Yield (Gallons / Dry Ton)	57.5	58.5	61.8	64.2	64.4	64.5	64.6	64.8	64.9
Total Fuel Yield (Gallons GE / Dry Ton)	51.3	52.2	59.1	61.4	61.5	61.6	61.7	61.8	61.9
Overall Carbon Efficiency to Hydrocarbon Products (% C in Feedstock)	28.2%	28.7%	29.9%	31.0%	31.0%	31.0%	31.1%	31.1%	31.2%
Water Consumption (Gal H2O / Gal C5+)	12.4	9.3	5.8	5.2	4.5	3.8	3.1	2.4	1.7
Electricity Export (kWh / GGE)	0	0	0	0	0	0	0	0	0

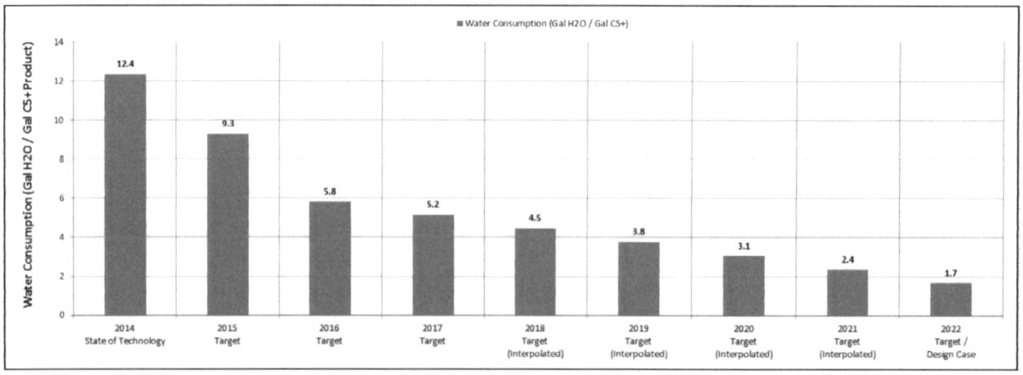

Product Quality Outlook for High-Octane Gasoline Pathway

Modeled product quality trends (3 of 3)

Product Quality Trends	2014 State of Technology	2015 Target	2016 Target	2017 Target	2018 Target (Interpolated)	2019 Target (Interpolated)	2020 Target (Interpolated)	2021 Target (Interpolated)	2022 Target / Design Case
Aromatics Content of C5+ Product (Wt%)	25.3%	15.0%	8.0%	4.0%	3.3%	2.6%	1.9%	1.2%	0.5%
C5+ Product Specific Gravity	0.75	0.73	0.71	0.70	0.70	0.70	0.70	0.70	0.70
C5+ Product API Gravity	58.17	62.41	68.57	69.37	69.52	69.68	69.83	69.98	70.14
Mixed C4 Product (Wt% Total Products)	27.2%	27.4%	0.0%	0.0%	0.0%	0.0%	0.0%	0.0%	0.0%
C5+ Aromatics (Wt% Total Products)	18.5%	10.9%	8.0%	4.0%	3.3%	2.6%	1.9%	1.2%	0.5%
C5+ Non-Aromatics (Wt% Total Products)	54.4%	61.7%	92.0%	96.0%	96.7%	97.4%	98.1%	98.8%	99.5%

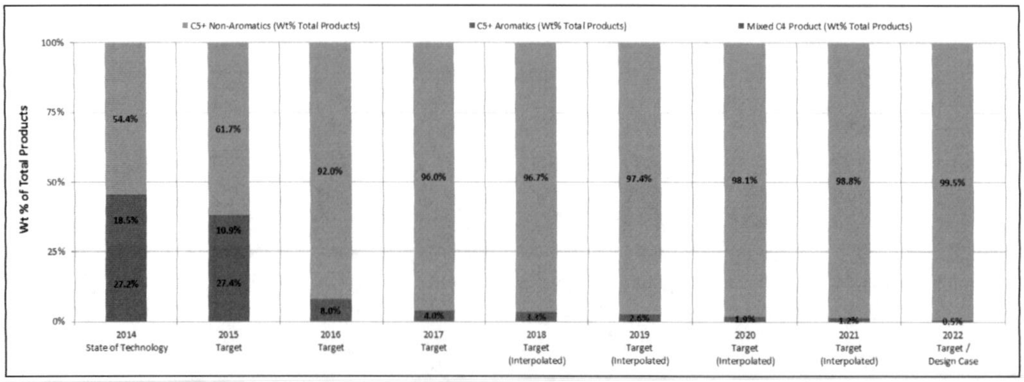

Acknowledgements

Michael Talmadge (NREL, co-author)
Abhijit Duta (NREL, co-author)
Jesse Hensley (NREL)
Jose Schaidle (NREL)
Mary Biddy (NREL)
Lesley Snowden-Swan (PNNL)

DOE's Bioenergy Technologies Office (BETO)
http://www.eere.energy.gov/biomass

Speaker information:
Eric.Tan@nrel.gov
http://www.nrel.gov/biomass

Assessing the Recovery of N-methyl-2-pyrrilodone from Waste Streams in the Specialty Chemical Industry

C. Stewart Slater, Mariano J. Savelski, Brigitte Pastore
Rowan University, Glassboro, NJ, USA

Frank A. Richetti, Nancy A. Uff
DuPont, Parlin, NJ, USA

Paper 752f
2015 Annual Meeting American Institute of Chemical Engineers
Salt Lake City, UT
November 2015

Rowan University

1

Introduction

- Case study is underway on waste reduction in specialty chemical manufacture
- EPA P2 Program – Region 2
- Partnership with DuPont plant in Parlin, NJ
  - Resins and resin precursors
  - Cyrel® printing products
  - ChromaPro® colorants
  - Teflon® and Autograph® coatings
- Case study focuses on resin precursor manufacture
- Batch-based process with relatively high solvent use
  - Focus on solvent reduction – parallels to Pharma industry

Rowan University

2

Background

- Polyimide (PI) and Polybenzoxazole (PBO) resin precursors
  - Applied and cured by DuPont's customers to form PI or PBO resins
- Resins essential to make
  - Microelectronic devices
  - Microelectromechanical systems
  - Aerospace components
- Synthesis requires solvents

A. J. Kirby, *Polyimides - Materials, Processing, and Applications*, Tarrytown, NY: Pergamon Press Inc., 1992.
TCS Electronics, Inc. *Assembly of Turn Flex Printed Circuit Board used within the Medical Industry* [Online]. Available: www.tcs-electronics.com
S. Leonard. (2010, May 10). *Making Sense of Medical MEMS Technologies* [Online]. Available: www.qmed.com

Rowan University

3

Background

- Solvent of interest in this study
  - N-Methyl-2-pyrrolidone (NMP)

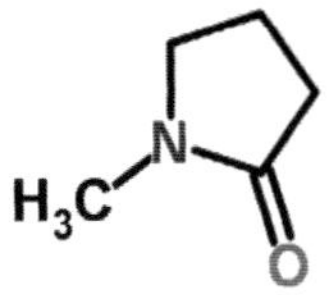

- Issues
  - Suspected reproductive and developmental toxin
  - Significant life cycle emissions from manufacture
  - Expense for purchase and disposal
- Beneficial properties
  - Dipolar aprotic solvent
  - High thermal stability
  - Produces high molecular weight resin precursor

Royal Society of Chemistry (2015). *N-Methylpyrrolidone* [Online]. Available: www.chemspider,com
M. Reisch, "Solvent Users Look To Replace NMP," *Chemical & Engineering News Archive*, p. 32, 2008.
Scientific Committee on Consumer Safety, "Opinion on N-Methyl-2-pyrrolidone (NMP)," European Union, 2011.

Data generated from SimaPro®

Rowan University

4

Background: Toxic Release Inventory

U.S. NMP Waste in 2013 by Sector

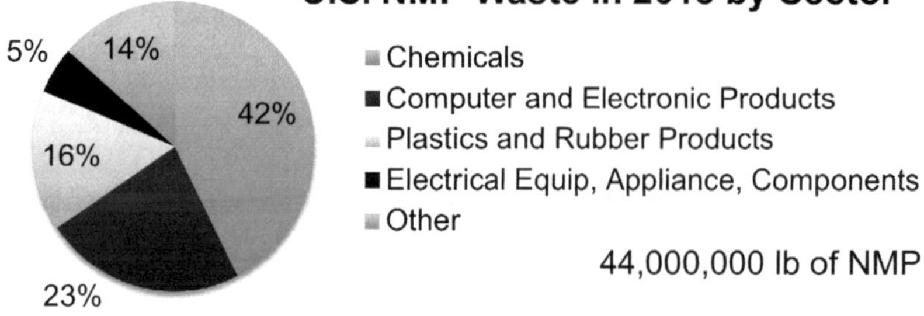

44,000,000 lb of NMP waste in 2013

Historical Trends in U.S. NMP Use

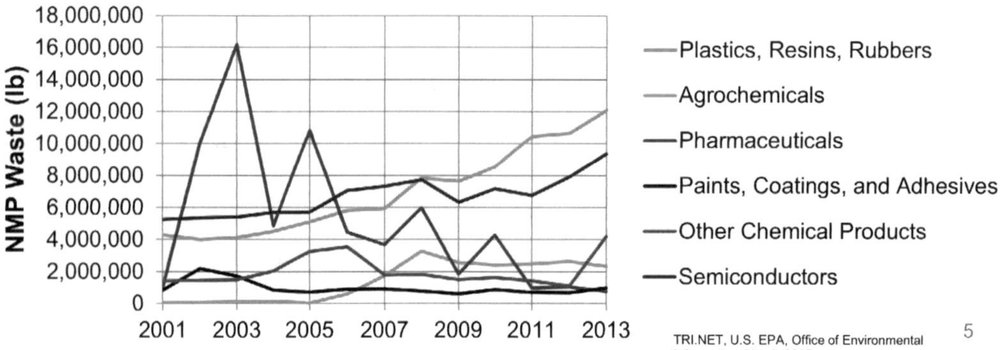

TRI.NET, U.S. EPA, Office of Environmental Information, March 2015.

5

Case Study Approach

- Evaluate current process
- Explore green engineering alternatives
  - Recover NMP from "hazardous" solvent waste
    - Reduce virgin NMP use and disposal
    - Design and analyze recovery system
  - Substitution of NMP with more benign solvents
  - Purify and reuse water from washwater waste
    - Design and analyze recovery system
- Analyze green engineering alternatives
  - Environmental and financial

Rowan University

6

Current Case

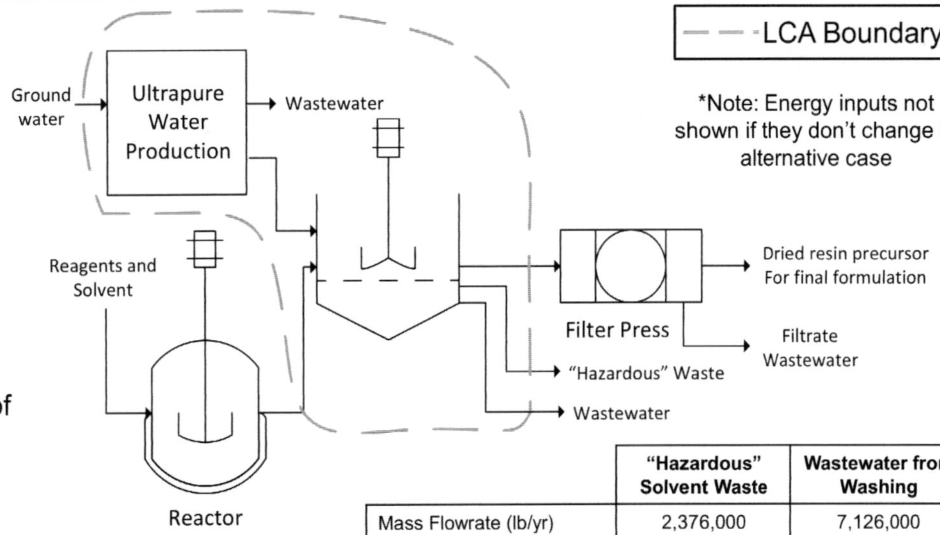

*Note: Energy inputs not shown if they don't change in alternative case

Basis: Annual amount of solvent used

Virgin NMP Flowrate (lb/yr)	404,000
Ground Water Flowrate (lb/yr)	15,535,000
Ultrapure Water Flowrate (lb/yr)	9,062,000
Ultrapure Water Production Wastewater Flowrate (lb/yr)	6,473,000

	"Hazardous" Solvent Waste	Wastewater from Washing
Mass Flowrate (lb/yr)	2,376,000	7,126,000
Water Composition (wt%)	81.5	99.85
NMP Composition (wt.%)	17	0.15
HEMA Composition (wt.%)	0.5	~50 ppm
HCl Composition (wt.%)	0.5	~50 ppm
TFA Composition (wt.%)	0.5	~50 ppm

Preliminary Analysis

- "First pass" approach examines separation of binary system of major components: NMP / H_2O
- Evaluate thermodynamics, concentration ranges, etc
- Evaluate different separation processes: distillation, pervaporation, etc.
- Screen alternative solvents
- Results: Distillation preferred method (non-azeotropic)

Rowan University

Proposed NMP Recovery

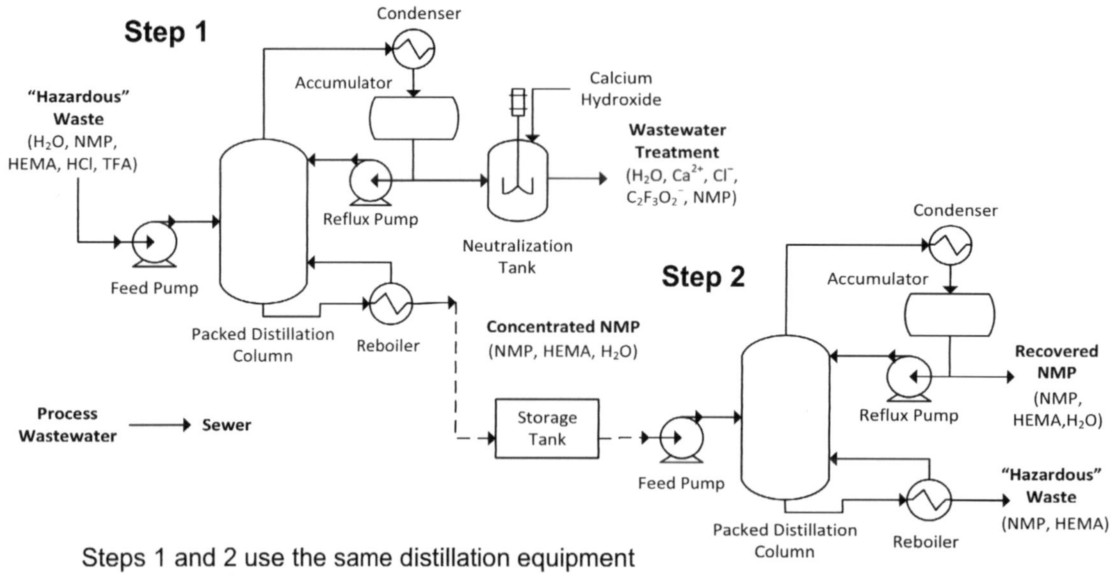

Steps 1 and 2 use the same distillation equipment operated in a sequential batch operation

Rowan University

9

Proposed NMP Recovery: Step 1

Step 1 operates 832 hr/yr

Calcium Hydroxide Slurry

	wt.%	Flow (lb/yr)
Water	85.0	90,000
Calcium Hydroxide	15	16,000
Total	100	106,000

"Hazardous" waste — from resin process

	wt.%	Flow (lb/yr)
Water	81.5	1,936,000
NMP	17	404,000
TFA	0.5	12,000
HCl	0.5	12,000
HEMA	0.5	12,000
Total	100	2,376,000

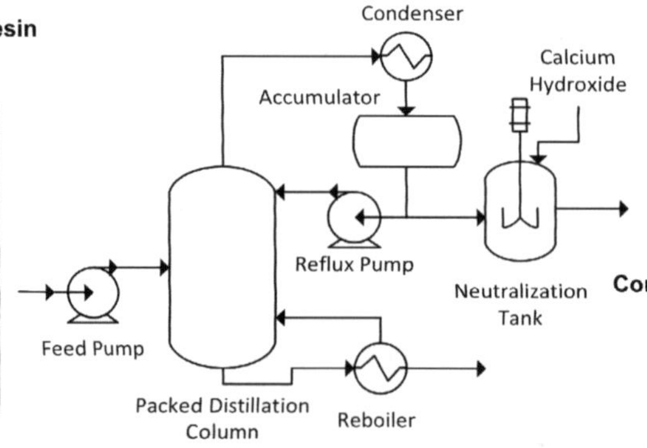

Distillate — neutralized and sent to wastewater treatment

	wt.%	Flow (lb/yr)
Water	98.36	2,034,000
NMP	0.05	1,000
HEMA	trace	trace
Ca^{2+}	0.44	9,000
Cl^-	0.58	12,000
$C_2F_3O_2^-$	0.58	12,000
Total	100	2,068,000

Concentrated NMP — sent to Step 2

	wt.%	Flow (lb/yr)
Water	0.03	100
NMP	97.11	402,000
TFA	trace	trace
HCl	trace	trace
HEMA	2.86	12,000
Total	100	414,000

Rowan University

10

Proposed NMP Recovery: Step 2

Step 2 operates 145 hr/yr

Concentrated NMP — from Step 1

	wt.%	Flow (lb/yr)
Water	0.03	100
NMP	97.11	402,000
HEMA	2.86	12,000
Total	100	414,000

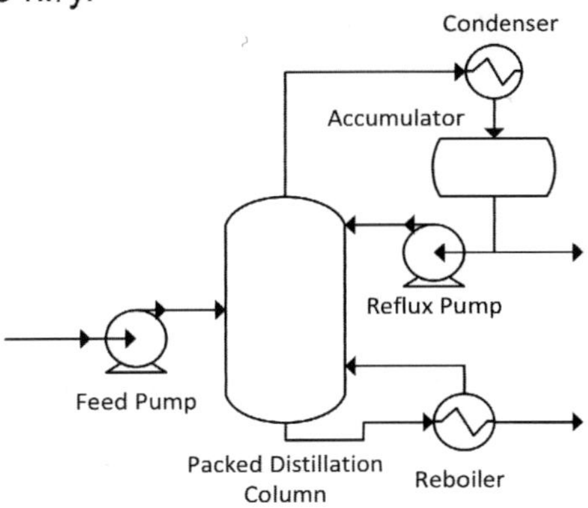

Recovered NMP

	wt.%	Flow (lb/yr)
Water	0.03	100
NMP	99.97	382,000
HEMA	trace	trace
Total	100	382,000

Bottoms — Disposed of as "Hazardous" Waste

	wt.%	Flow (lb/yr)
Water	trace	trace
NMP	62.96	20,000
HEMA	37.04	12,000
Total	100	32,000

This process recovers 95% NMP at 99.97% purity

Rowan University

11

Method of Analysis

- SimaPro®
  - Life Cycle Assessment
    - Total Emissions
    - ReCiPe
- AspenTech®
  - Process Design
  - Economic Evaluation
- R.SWEET
  - Software tool developed by Rowan

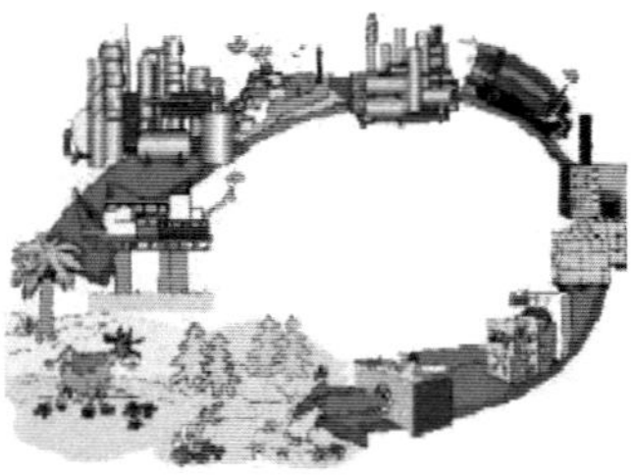

aspentech

Rowan University

E. J. Cavanagh, M. J. Savelski and C. Slater, "Optimization of environmental impact reduction and economic feasibility of solvent waste recovery using a new software tool," *Chemical Engineering Research and Design*, vol. 92, no. 10, pp. 1942-1954, 2014.

12

ReCiPe Damage Assessment

- Transforms individual life cycle category results into limited number of individual scores
- ReCiPe (H/A) — hierarchist approach, most common
- Scores express relative severity on an impact category: 18 Midpoints are combined into 3 Endpoints
  - Human Health, Ecosystems, Resources
- Endpoints are normalized and weighted to produce a single score
  - Human Health and Ecosystems weighted by 400 Pt
  - Resources weighted by 200 Pt

Rowan University

www.lcia-recipe.net

M. Goedkoop, R. Heijungs, M. Huijbregts, A. De Schryver, J. Struijs and R. van Zelm, ReCiPe 2008, 2009.

13

Life Cycle Inventories and Costs

Manufacturing ("Cradle") LCIs, Damage factors, and Costs for Raw Materials
(1 lb basis)

	NMP	Ultrapure Water	HEMA*	HCl	TFA*	$Ca(OH)_2$
Total Emissions (lb)	4.22	0.0115	5.78	1.50	4.80	0.751
CO_2 (lb)	3.72	0.00913	5.59	1.27	4.16	0.740
Human Health (mPt)	66.2	0.143	105	31.9	112	10.3
Ecosystems (mPt)	34.9	0.0781	55.4	11.9	40.0	6.28
Resources (mPt)	97.7	0.104	148	22.2	92.6	4.98
Purchasing Cost ($)	1.95[a]	0.00613[a]	1.06[b]	0.126[b]	0.126[b]	0.113[c]

*Required chemical tree or substitution technique to estimate LCI

[a]Provided by DuPont
[b]ICIS (2008). *Indicative Chemical Prices A-Z* [Online]. Available: http://www.icis.com/chemicals/channel-info-chemicals-a-z/
[c]M.C. Schrock and J.K. Mercer, "Types of Calcium Reagents and Their Application in Acid Mine Drainage and Abondoned Mine Land Reclamation," *National Meeting of the American Society of Mining and Reclamation*, Pittsburgh, PA, 2010.

Data generated from SimaPro®

Rowan University

14

Life Cycle Inventories and Costs

LCIs, Damage factors, and Costs for Utility Production and Waste Disposal ("Grave")

	Utilities (1 MJ basis)		Waste Disposal (1 lb basis)	
	Electricity	HP Steam	"Hazardous" Waste	Wastewater
Total Emissions (lb)	0.245	0.148	0.217	0.0280
CO_2 (lb)	0.211	0.147	0.193	0.0275
Human Health (mPt)	4.26	2.01	2.52	0.404
Ecosystems (mPt)	1.91	1.26	1.62	0.235
Resources (mPt)	4.78	3.38	-3.95	0.103
Cost ($)	0.0314[a]	0.00682[b]	0.241[c]	0.00475[c]

Hazardous solvent waste sent to off-site incinerator
Wastewater from water washing and recovery distillation step 1 is sent to publically–owned treatment works (POTW)

[a]US Energy Information Administration (2015). *Electricity Data Browser* [Online]. Available: http://www.eia.gov/electricity/
[b]G. Towler and R. Sinnott, *Chemical Engineering Design*, New York: Butterworth-Heinemann, 2013.
[c]Provided by DuPont

Rowan University

Data generated from SimaPro®

15

Life Cycle Emissions

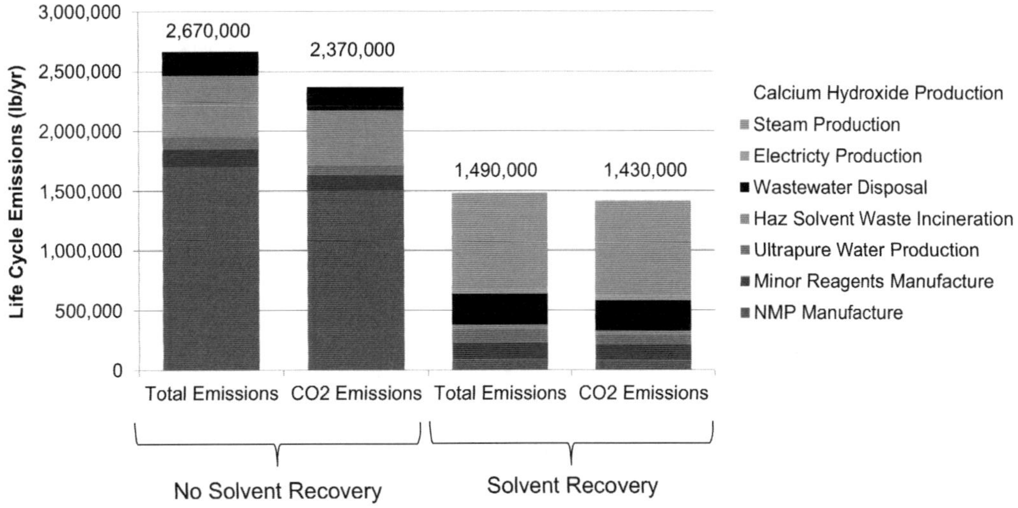

44% Reduction in Total Emissions
40% Reduction in CO_2 Emissions

Rowan University

16

ReCiPe Damage Assessment

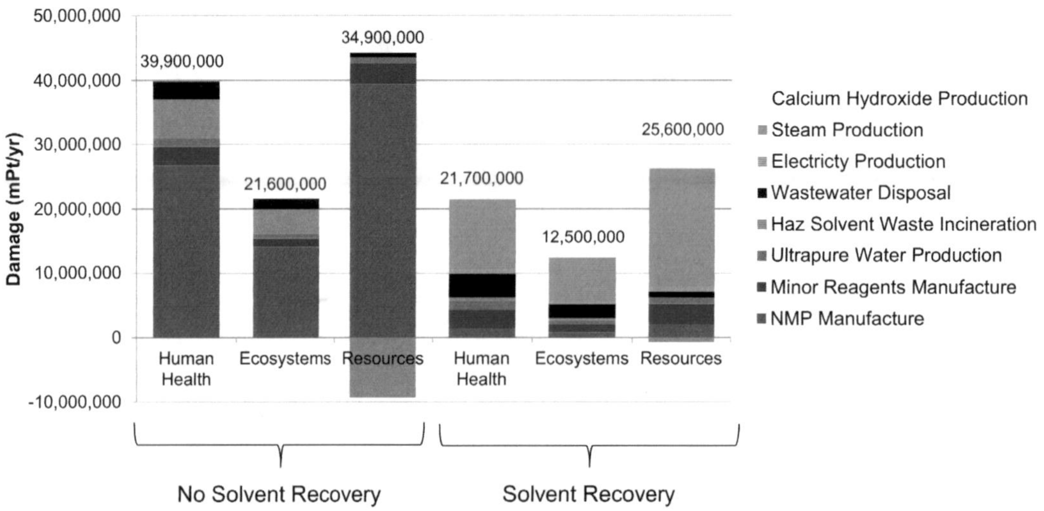

46% Reduction in Human Health Damage
42% Reduction in Ecosystems Damage
27% Reduction in Resources Damage
38% Reduction in Total Damage

Rowan University

17

Economic Life Cycle Assessment

Operating Costs

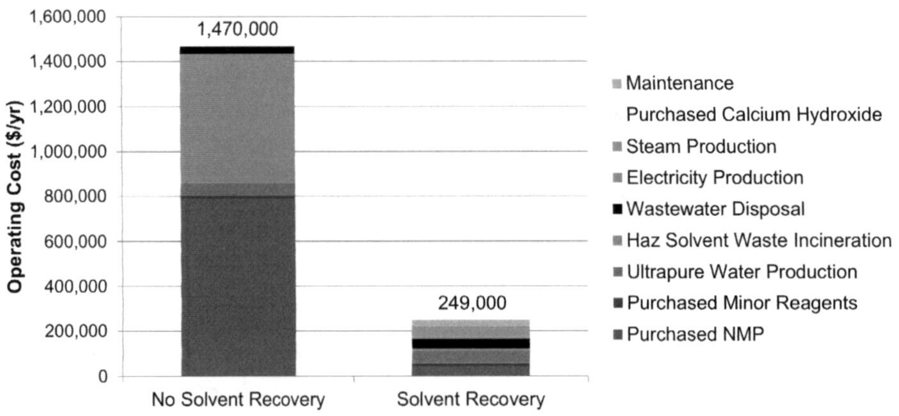

83% Annual Cost Savings

Rowan University

18

Alternative Solvents

- **DMSO** and **sulfolane** are favorable substitutes
  - Less hazardous
  - Reduced environmental impact
- Solvent substitution analysis
  - Substituted at a 1:1 ratio
  - Distillation recovery design varies for each solvent

Manufacturing LCIs and Costs for Solvents (1 lb basis)

	NMP	DMSO	Sulfolane*
Total Emissions (lb)	4.22	1.31	1.10
CO_2 (lb)	3.72	1.18	0.749
Human Health (mPt)	66.2	21.8	29.3
Ecosystems (mPt)	34.9	11.8	7.2
Resources (mPt)	97.7	54.9	44.3
Purchasing Cost ($)	1.95[a]	2.10[b]	3.09[b]

*Required chemical tree technique to estimate LCI

Data generated from SimaPro®

[a]Provided by DuPont
[b]"ACS Grade Solvents," Spectrum Chemical MGF Corp, 2015. [Online]. Available: https://www.spectrumchemical.com/

Rowan University

U. Tilstam "Sulfolane: A Versitle Dipolar Aprotic Solvent," *Org. Process Res. Dev.*, vol. 16, pp. 1273-1278, 2012.
A. Figoli *et al.* "Towards non-toxic solvents for membrane preparation: a review," *Green Chemistry*, vol. 16, pp. 4034-4059, 2014.

19

Life Cycle Emissions

- NMP, DMSO, and sulfolane are high boiling solvents – high pressure steam is needed for recovery
- DMSO and sulfolane recovery designs require vacuum distillation
- Sulfolane is recovered in 1 step - b.p. of sulfolane > b.p. of HEMA

Reduction in Annual Total Emissions
DMSO Substitution = 44%
Sulfolane Substitution = 47%
NMP Recovery = 44%
DMSO Substitution with Recovery = 45%
Sulfolane Substitution with Recovery = 50%

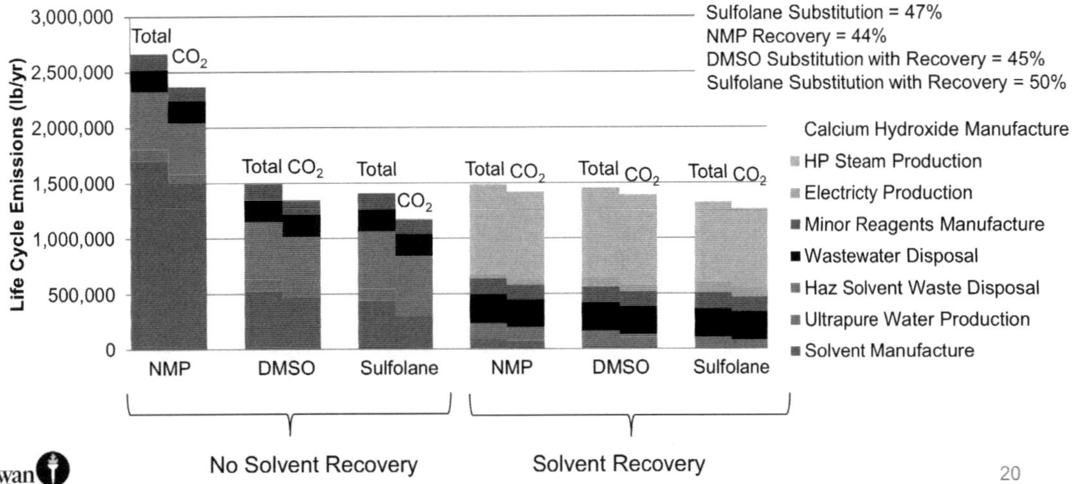

Rowan University

20

Economic Life Cycle Assessment

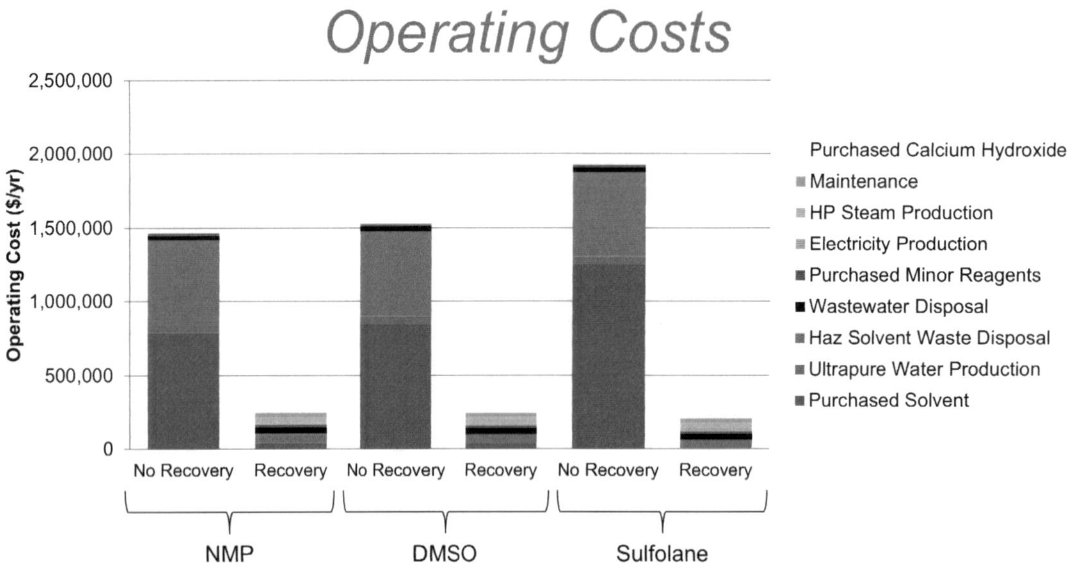

Reduction in Annual Operating Costs
DMSO Substitution = -4%
Sulfolane Substitution = -31%
NMP Recovery = 83%
DMSO Substitution with Recovery = 83%
Sulfolane Substitution with Recovery = 86%

Rowan University

21

Solvent Recovery Economics

Installed Capital Cost of Solvent Recovery Equipment (Distillation)

Solvent	Capital Cost ($)
NMP	1,497,000
DMSO	1,480,000
Sulfolane	1,566,000

Projected Economic Results for Solvent Recovery

Solvent	ROI	IRR	Payback Time (after tax)	NPV at 5 yr	NPV at 10 yr
NMP	59%	67%	2.2 yr	$1,750,000	$3,128,000
DMSO	60%	69%	2.2 yr	$1,781,000	$3,165,000
Sulfolane	58%	66%	2.3 yr	$1,790,000	$3,214,000

Notes: 7 year MACRS depreciation method
35% Tax Rate
15% Interest Rate

Rowan University

22

Overall Analysis of Alternatives

Scenario	Total Emission Reduction	CO_2 Emission Reduction	Damage Assessment Reduction	NPV at 10 yr
DMSO substitution without recovery	44%	43%	46%	$-298,000
Sulfolane substitution without recovery	47%	51%	50%	$-2,306,000
▶ NMP recovery	44%	40%	38%	$3,128,000
DMSO substitution with recovery	45%	41%	39%	$3,165,000
Sulfolane substitution with recovery	50%	46%	44%	$3,214,000

- Solvent substitutes appear slightly better in environmental categories, but don't yield significant economic improvement
- Recommended approach — NMP with recovery
  - If purity is sufficient, avoids validating alternative solvents

Rowan University

23

Other Processing Options

- NMP recovery without capital expenditure – assuming distillation equipment exists elsewhere
  - Transportation cost to ship NMP waste to alternative plant or Solvent Recovery facility
  - "Toll" cost to reprocess – will be higher if sent to outside Solvent Recovery facility

  - Transportation cost to ship purified NMP back to plant or for 3rd party use
- Water recovery from wastewater - production of purified water
  - Limited potential impact on LCA and cost benefit

Rowan University

www.cleanharbors.com

24

Summary

- Solvent recovery is a viable option for manufacturing platforms in the specialty chemical industry
- Case study shows potential benefits in $ saved and emissions reduced if a recovery scheme can be implemented
- Much more attractive if recovery equipment (distillation) capacity is available
- Future work
  - Low-energy separations
  - Alternative solvents
  - Reach out to other plant locations
  - Water reuse evaluation/design and LCA
  - Reverse osmosis / Pervaporation / Adsorption / IX for washwater recovery

Rowan University

www.ddpsltd.co.uk/distillation-and-evaporation,49910/

25

Acknowledgements

- Support for this research is provided by a grant from the U.S. Environmental Protection Agency – Region 2 - NP-96291314-0
- We acknowledge the support of DuPont Corporation and Michael Hatton Jr, Charles Richwine, Jeffrey Staniszewski (ret) of DuPont
- We acknowledge assistance of the following students: Matt Buchanan, Ben Ficara, Joe Garrett, Steve Hertzog, Ryan Liberty, Kyle Surette, Christian Wisniewski

Rowan University

26

THANK YOU
FOR YOUR INTEREST
AND BEING HERE
THURSDAY PM

THANK YOU TO THE
SESSION CHAIRS

Rowan University

27

431372 Looking into the Future of the Ethylene Industry: A Generic Assessment Model for Emerging Technologies

Abstract:

Ethylene is the largest volume chemical derived from natural gas liquids (NGL)[1]. Recent shale gas boom leads to decreasing price of NGL and natural gas, and as a result boosts its production and exports[2]. At the same time, ethylene is one of the largest energy consumer and GHG emissions resources in the chemical industry. Based on IEA's analysis, ethylene production accounts for 13% of energy use and 15% of GHG emissions for global chemical industry[3, 4]. Thus the rapid growth of ethylene production adds considerable amount of energy use and GHG emissions to the chemical industry. A lot of efforts have been taken to develop energy efficient technologies for the ethylene production, and it is critical to assess the potential life cycle impacts of them on energy saving and GHG emissions mitigation over temporal and geographic scale. However, due to lack of inventory data, it is difficult to conduct such an assessment for new technologies at a economy-wide scale, especially for those in their early stages.

In this work, the impacts of several emerging technologies that aim to improve energy efficiency of ethylene production are evaluated by a generic modeling framework. This model integrates chemical process design and modeling, life cycle assessment, techno-economic analysis and scenario analysis. A brief introduction of this model will be given in the beginning of the presentation. Then a case study of ethylene industry will be presented in the details. First, the conventional technology module in the modeling framework is used to model all direct and indirect unit processes involved in current ethylene production life cycle. Second, new technology module is activated to assess the net impacts of emerging technologies on energy use and GHG emissions. These impacts are compared with the convention technologies to shed light on improving opportunities in a product level. Last, the results of both conventional and emerging technologies are integrated with scenario analysis module to reflect the energy, emissions and economic implications of technology changes for the national-level ethylene production system in future decades. This model provides a reliable methodology to overcome knowledge barriers and deliver robust insights of potential influences that emerging technologies can have at a macro-level from life cycle perspective. The results can provide policy makers with good references on decision making for technology investment and promotion, offer manufactures with good understanding of key drivers of energy, emissions and resources use for a certain production system, provide researchers with insightful information on the possible bottleneck and future direction of R&D.

References

[1] B. Fallas and P. Pavlov, "Platts special report: petrochemicals, time to get cracking " McGraw-Hill Companies, Inc. , New York, NY, USA2013.

[2] ACC, "Year-end 2013 chemical industry situation and outlook," American Chemistry Council Washington, DC, USA2013.

[3] Y. Yao, D. Graziano, M. Riddle, J. Cresko, and E. Masanet, "Greener pathways for energy-intensive commodity chemicals: opportunities and challenges," *Current Opinion in Chemical Engineering,* vol. 6, pp. 90-98, 11// 2014.

[4] IEA, ICCA, and DECHEMA, "Technology Roadmap, Energy and GHG Reductions in the Chemical Industry via Catalytic Process," IEA, ICCA, DECHEMA, France2013.

AUTHOR INDEX

AUTHOR INDEX

AUTHOR INDEX

AIChE
120 Wall Street, FL 23
New York, NY 10005-4020

ISBN 978-1-5108-1852-1

CURRAN ASSOCIATES INC.
proceedings
.com